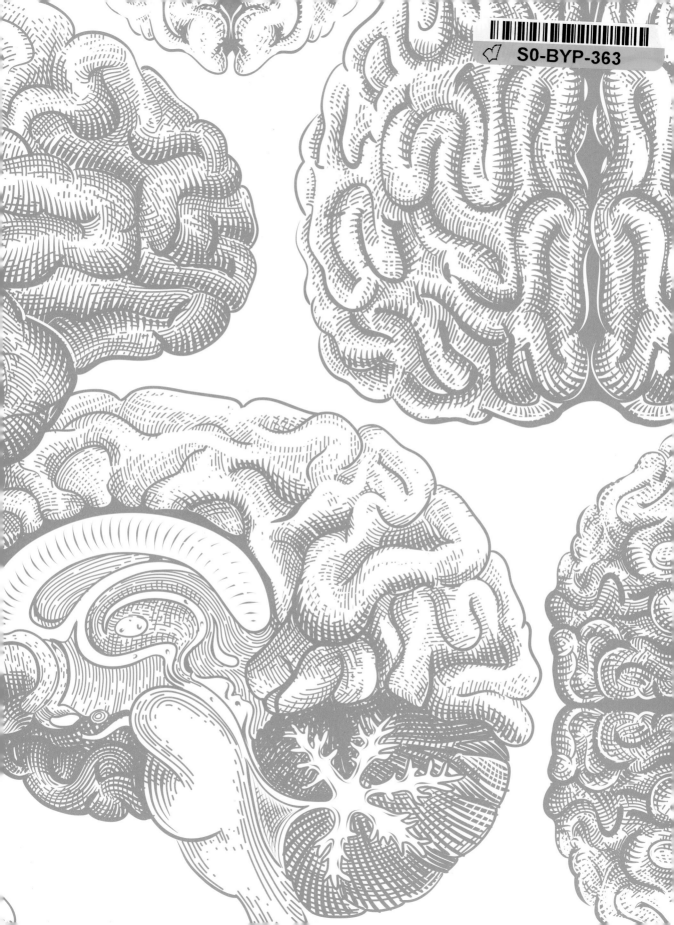

IT'S ALL IN YOUR HEAD

BRAIN STORMS, GRAY MATTER, AND WHAT MAKES YOU *YOU*

by **KEITH BLANCHARD**

WICKED COW STUDIOS

WICKED COW STUDIOS

45 West 21st Street, Floor Two • New York, NY 10010

Wicked Cow Studios
President/CEO: Michael Hermann
COO: Chris Flannery
SVP/General Counsel: Elliot Schaeffer
VP/Publishing: Frank Fochetta
Associate Editor: Mallory Stratton
Associate Franchise Manager: Matt Russo
Creative Director: Samantha Merley
Finance: Lenny Sander, Bill Pellegrino, Mike Sander, Ben Sander, Jimmy Ryan, Gina DiRubba, Gale Patterson, and Haydee De Jesus (The Creative Planners Group, Ltd.)

It's All In Your Head Contributors
Publisher/Executive Editor: Michael Hermann
Author/Editor: Keith Blanchard
Publishing Operations: Frank Fochetta
Creative Director/Book Design: Samantha Merley
Associate Editor: Mallory Stratton
Associate Producer: Matt Russo

Based on an original idea by: Adam D. Hirsch
Original cover illustration: Michael Morgenstern

Special Contributors
Producer: Jennifer Fistere
Senior Researcher: Roxanne Palmer
Line Editor: Lisa Catherine Harper
Senior Consultants: Chris Flannery, Richard Taub, Lou Wallach
Special Photo Credits: Don Hermann, Christopher Lynch, Snorri Sturluson, Stephen Gill, Michael Hermann
Infographic illustrator: Josh McKible
Special Art: Julie Rossman
Indexer: Kitty Chibnik
Additional Graphic Art and Design: Kip Helton, Matt Cokeley
Wicked Cow Studios Logo: Joseph Ari Aloi
Junior Researchers: Austin Day, Jeremy Cohen, Nina Levitin, Tyler Ackerson, Emma Duncliffe, Carolina Cepeda, Haley Muratore, Nick Servidio, Ethan Messinger

Printed in China

Library of Congress Control Number 2017948042
ISBN Number 978-0-692-91823-4

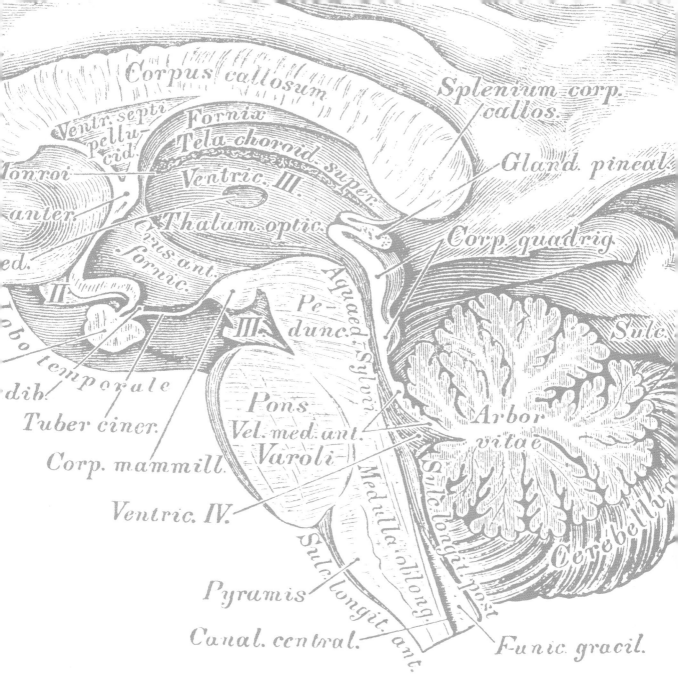

Many Special Thanks

Neil deGrasse Tyson, Helen Fisher, Jason Harrison, Dana Harrison, Bill Campbell, Peter Berkowitz, Michael Schrage, Joe Meli, Carmin Romanelli, Kevin Foley, Bridget Fitzgerald, Ann Amstutz Hayes, Kristy Hammam, Allison Wallach, Harrison Land, Stephen Land, Billy Goldberg, Rob Russo, Andy Gaies, Tom Hayden, Jeff Frumin, Jeff Feinstein, Mark Bauman, Dave Blum, Kostya Kennedy, Megan Pearlman, Alyssa Smith, Damian Slattery, Lauren Moriarty, Steve Geppi, Joe Foss, Chris Powell, Jeff Vaughn, Kuo-Yu Liang, Tim Lenaghan, Elaine Liu, Marty Ordman, Megan Kingery, Paul Woodruff, Eric Hurwitz, Stephen Colvin, Gerry Ohrstrom, Heather Greenfeig, Michael Breach, Lisa Johnson

HERMANN'S HEAD

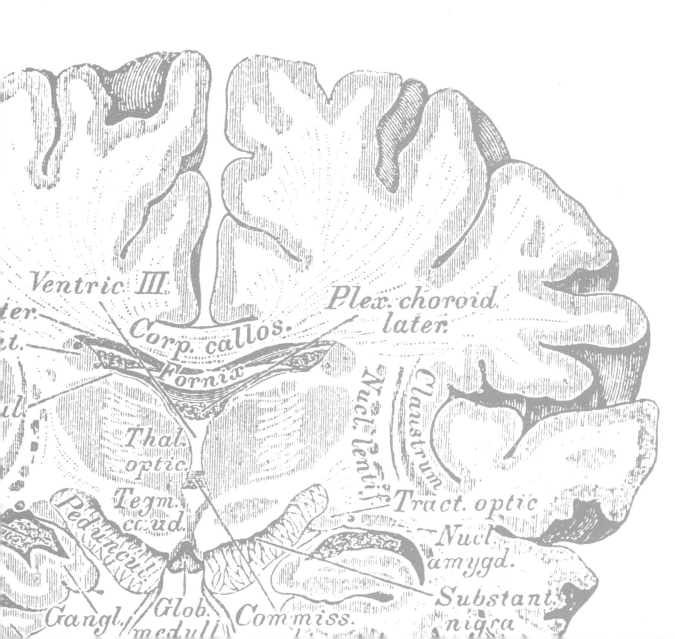

Ventric III.

Plex. choroid.
later.

Corp. callos.

Fornix

Nucl. lenti?

Claustrum

Thal.
optic.

Tract. optic

Tegm.
caud.

Nucl.
amygd.

Pedunculi

Substant.
nigra

Gangl. Glob. Commiss.
medull

If I had my way, I wouldn't be writing this. I wouldn't have spent the past two years building yet another new company on the back of yet another crazy idea. But I don't make the decisions around here, my brain does. And my brain craves the rush of the new, whether it's an idea, an opportunity, a friend, an experience. If my brain doesn't get its regular fix, it becomes dissatisfied and starts poking at me and tormenting me. So here I am now, writing the introduction to a book that's all about the thing that is making me write the introduction to the book.

But a funny thing happened along the process of creating it. As our team began unwinding—and getting tangled up in—all the many mysteries of the human mind, I began to see more clearly why I'm driven to do things that cause me so much anxiety, risk, agita...and reward. It turns out that I'm an incorrigible entrepreneur because my brain is simply wired that way.

So the bad news is, I really have no choice but to take leaps into the unknown, to explore this mental frontier, to try to build a business from it. The good news is, I'm more excited than I've ever been about an endeavor. The human mind is still so unknown, so mysterious, So... new. There's enough unexplored territory in the mind to keep my brain happy and stimulated for a long, long time.

Michael Hermann
Publisher/Executive Editor
June 28, 2017

Dedicated to my love, Stacy. And, of course to my Mom, Dad, and to you, Lisa. You're always with me.

I love what's in your heads: Keith, Sam, Mal, Matt, Adam, Frank, Flannery, Elliot. Thank you for your unbridled efforts, support, and talents.

Emotional Advisory Committee
Lois Schwartzman, Chris Golier, Rich Taub, Christopher Lynch, Ahovi Kponou, Chris Mack, Paul Hough, Danielle Wayne, Joe Cohen, Inci Ulgur, Leslie Blanchard, Rich Ticknor, Rita Ticknor, Michelle Ticknor, Matt Ticknor, Brandi Mayo, Rachel Hanfling, Doug Rose, Greg Werner, Stephen Gill, Scott Roskind, Russell Thomas, Peter Haugen, Nathan Grouse, Jo Hackett, Steve Hackett, Daymond John, Ted Kingsbery, Eric Simon, Jason Flom, Faith Wall, Snorri Sturluson, Jill Zarensky, Rob Nelson, Yasheika Harris, Perrie Briskin, Julie Sharbutt, Regina Servidio, Carrie Capstick, James Berman, Fran Schaeffer, Catherine Flannery, Kathy Fochetta, David Gilmour, David J. Mullany, Stephen Mullany, David A. Mullany, Kristi Vilberg, Malcolm Jones, Lauren Sher

Junior Committee
Genevieve Taub, Blake Cohen, Maddie Cohen, Sasha Berman, Hope Harrison, Eli Harrison, Ben Harrison, Zeki Hirsch, Tommy Umano, Mathew Umano, Ryan Umano, Cole Umano, Brendan Umano, Alex Umano, Christopher Umano, Ethan Coats, Ellis Coats, Brennan Taylor, Harrison Taylor, Creighton Taylor, "Free" Grady Golier, Christopher Golier, Elizabeth Golier, Charley Hough, London Hough, Will Flannery, Ava Flannery, Celia Flannery, Samantha Schaeffer, Russell Schaeffer, Marla Schaeffer, Bo Hanfling, Max Hanfling, Chloe Blanchard, Sam Blanchard, Jonah Blanchard, Jake Fochetta, Mateo Fochetta, Cameron Stratton, Mackenzie Stratton, Paige Russo, Leo Pickering, Nimo Pickering, Mika DiDomenico, Will Strelinger, Chloe Werner, Eden Werner, Holden Werner, Gaston "Jerry" Jones

CONTENTS

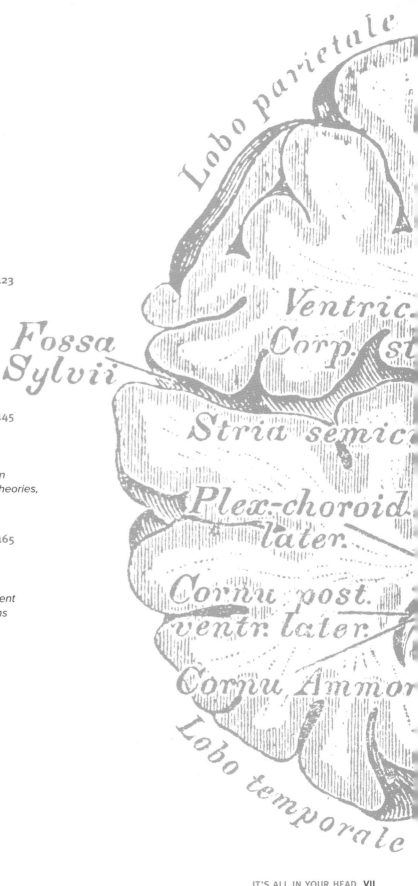

INTRODUCTION

Last year my father, a semi-retired construction worker, stepped up into the back of his truck like he'd done ten thousand times before, and missed his footing. He was alone in his own driveway, set to begin a day of work at home, so when he slipped and fell backwards out of that van, and cracked the back of his head on the Belgian-block curb, and blacked out, in a widening pool of blood, there was nobody to see it, nobody to call an ambulance.

He lay there for hours, until he finally awoke, flat on his back and blinking in the overhead sun, and fumbled for his cellphone to call my mother.

The damage was done. A routine concussion check failed to reveal that the precision of my dad's fall against the corner of the curb had initiated a set of seizures deep inside his brain. After a few weeks he started losing specific memories; shortly whole years of his life were wiped out. By the time he got the specialist help he needed—and the antiseizure medicine that prevents further damage—most of his memory was gone. Today new memories stick, but his 70-year lifetime is a patchwork of disassociated details, contextualized as far as possible by his very patient wife of 50 years. Dad doesn't remember high school; he remembers what my mom has told him about his high school, which in turn is only that subset of what she remembers him telling her over the years.

Will you indulge me for a few seconds?

Put your palms together as if you're praying. Then curl your fingers so your knuckles are touching, thumbs pressed against the side of your curled index fingers. The church without the steeple, if you will. Got it?

OK, now look at your fists: This is your brain. Your personal computer, your preferences and memories, your personality and history, your dreams and beliefs and desires. Everything you are and will ever be.

Does it look small? It is; it is. Your brain weighs about the same as a bottle of wine. But the incredible electrochemical calculating machine it houses is transcendently big—functionally infinite. From blobs just like this emerged the works of Shakespeare and Bach, the cure for polio, the space program. Most of your brain is actually a sheet, interestingly enough: just a few millimeters thick and folded in on itself

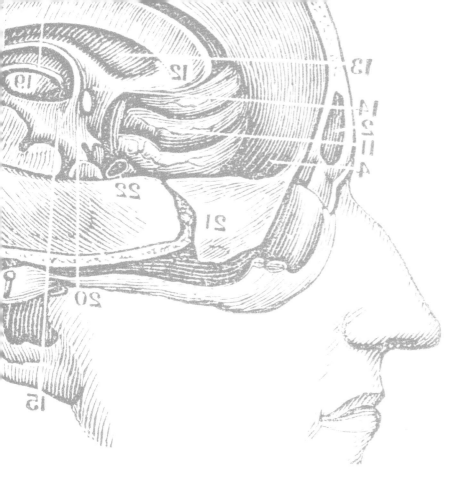

and stuffed into your brain like organic origami. But this malleable sheet contains billions upon billions of finely gradated decision gates, united by trillions of ever-morphing connections, all of it electrochemically firing in complex, not-precisely-repeatable sequences that we experience as thoughts, about everything you've ever known, dreamed, loved or hated or feared or trusted.

Your brain is not the same as mine, and yet we can communicate, share ideas, pledge partnership or enmity, agree to disagree. And the human mind, the software built into this hardware, somehow pulls all these discrete "thoughts" and impressions into actionable insights that let you react with lightning speed: to remove your hand from the hot stove, or to defend against a hungry leopard or slick salesman. Further, it unites your history and preferences into identity, and conveys to you a sense of individual self that so far seems to

be unique in the impulse-driven animal kingdom. The kitten assumes a similar kitten lives inside the mirror, but you do not—in fact you can draw humor from the scene, and can videotape it to share with other smugly superior humans, as the hapless kitten puffs up and arches its back and bounces hilariously into and away from its reflection.

I'm not a scientist—that's important. I'm just an insatiably curious bipedal primate who has spent fifty years trying to understand everything I can before I die. I've always been intrigued by challenging complexity and drawn to try explaining it to others, and I've occasionally received training and even compensation in the service of this, which is maybe a rough definition of journalism. Over the years I've created, and led teams of likeminded creative, for the likes of Maxim, Rolling Stone, The Wall Street Journal, Mental Floss, and a number of other places. Most recently I was Chief Digital Officer

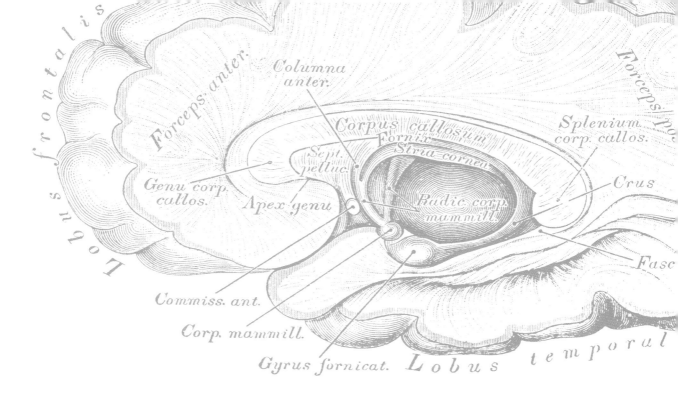

INTRODUCTION (CONT'D)

at the World Science Festival, where I first came in close contact with some of the world's top scientists and began to understand just how fast the frontier of neuroscience is moving.

And so, being a proud brain owner myself, I decided to dive in deep. I've spent the last several months investigating science's rapidly developing understanding of how our brains and minds work. I've spoken to scientists and researchers, journalists and academics, musicians and mathematicians and poker champions. And I can tell you, from my folding chair here at the frontier, it's fascinating to watch it begin to tie together. New brain imaging tools are literally allowing us to read minds; new medical tech is beginning to cure blindness and paralysis; new research is exposing just how faulty our memory and sensation actually are (a hint as to why magic and illusions work on us).

It's a humbling experience. We walk around as if our memory is photographic, our senses reliable, and our beliefs facts, but all of that

confidence is misplaced—it's simply not what the brain does. I've learned to think of my brain not as a reliable supercomputer, but rather as a super fast approximation machine that is almost never exactly correct, but is almost right in a way that's usually quite useful. It can alert us to the possibly dangerous and the probably edible. It can reassure us, in an inexact way, that we may have been here before; it can suggest whether or not to trust this car salesman. And it's fragile, despite its hard-won evolutionary helmet. Anything that can go wrong does go wrong, given enough people.

But for all its weaknesses on the reliability front, the brain is also remarkably plastic and changeable, and thousands of people are working furious overtime hours expanding the boundaries of what it can do. We're curing—or coming to grips with—diseases and conditions that once hobbled us. Building a world where all objects can talk to one another, granting access to the unthinkably vast resource of a connected world. We're approximating, and

hoping to supersede, the limits of our brain-power with artificial intelligence. This isn't just a fad to be amusingly considered anymore—it's an evolutionary leap into the unknown, a cliff edge we are running toward, come what may, at a high rate of speed. It has long been prophesied that The Singularity is coming, a moment when our machines will become smarter than we are. Well, surprise: It's almost certainly going to happen in your lifetime, and it's unclear how or even if humanity will survive it in any meaningful fashion. (Our science fiction is not bullish.)

We're even looking into beating back death itself, with all manner of deranged hail-mary experiments: head transplants, creating brand-new brains in the lab, "uploading" your brain's contents into some sort of digital storage. As

You may initially just think of yourself as "reading," but I hope by the end of the book you'll recognize what's actually happening. Your eyes will convert patterns of reflected light into electrochemical signals that will fire through the middle of your head, cascading into patterns in language recognition centers, and storing as memory. As the unconscious part of your brain plods along, managing core functions like breathing and heartbeat and temperature regulation, your brain will quickly recognize "words" and "sentences" and start comparing them to similar phenomena you've heard or read before. Higher level experiences will emerge: You'll understand phrases and sentences, decide whether you agree, judge whether my writing style is good or pedantic, my points valid or dopey, my meanderings worth sharing or a waste of your time. You may

It seems increasingly likely that yours will either be the first generation to live forever, or the last generation to die. The difference, of course, is everything.

crazy as it sounds, it seems increasingly likely that yours will either be the first generation to live forever, or the last generation to die. The difference, of course, is everything.

The brain's a complex place, and I've tried to organize this book in a rough chronology, from the simplest understanding how the physical brain is organized and how thoughts work, to how the senses feed information to the brain and how memories form, through the phenomena of consciousness like sleep and dreaming, and so on, right through to the future of artificial intelligence and the big "what's next." Along the way, we'll take a look at what can go wrong with the brain, how genius and creativity and madness are related, where belief and spirituality and sexuality come from.

be inspired creatively, or think of a loved one; you may drift off during the boring parts. (Sorry about that.)

But whatever happens, this much is sure: Your experience of reading the book will be completely and emphatically different than everybody else's, because the mind you bring to the table—your filter—is unique to you. In a very real sense, this is your book, and yours alone. So: Hello, it's nice to meet you, and I hope you have a good time. We've got a lot of work to do, so let's get started.

Keith Blanchard
October 2016, Chatham, NJ

YOUR NOODLE IN A NUTSHELL

COLOR CODED!

You may remember from seventh grade bio class that four key components of your neurological system are neurons, axons, synapses, and dendrites. **Neurons** are like little trees, with multiple outstretched branches called **dendrites** and a single long trunk called an **axon**. The axon communicates information out to other neurons' dendrites. The junctions between the axon of one neuron and the dendrite of another are called **synapses**, and it's across these tiny gaps, the **synaptic cleft**, that communication, at its most basic level, takes place.

MEET YOUR **NEUROTRANSMITTERS**

A few of the important chemicals coursing through your mind, and what they do

GLUTAMATE The most common neurochemical is associated with learning and memory. Too much can lead to impulsive behavior/violence.

GABA If glutamate is the throttle, GABA is the brake; it increases tranquility and makes sure you don't get too aggressive.

OXYTOCIN This "love drug" makes you fall in love and feel connected; helps nursing mothers bond with their babies.

DOPAMINE The "reward" chemical associated with pleasure of many kinds. Too little can lead to depression; too much can lead to addiction.

SEROTONIN Associated with a general peacefulness, serenity, and hopefulness. Low levels can lead to depression.

ADENOSINE This neurotransmitter builds up throughout the day; when enough has accumulated, you will feel sleepy.

ENDORPHINS These are your body's painkillers. At high levels you get relaxed and euphoric; at low levels you may be more sensitive to pain.

NORADRENALINE Like its cousin "adrenaline," noradrenaline heightens energy, and is involved with fight or flight and elevating heart rate.

Corpus Callosum is a nerve bundle connecting your brain's right and left hemispheres

Basal Ganglia helps planning and coordinating physical activity

Midbrain processes rudimentary vision and hearing

Pons is involved in breathing, sleeping, swallowing, and similar involuntary activities

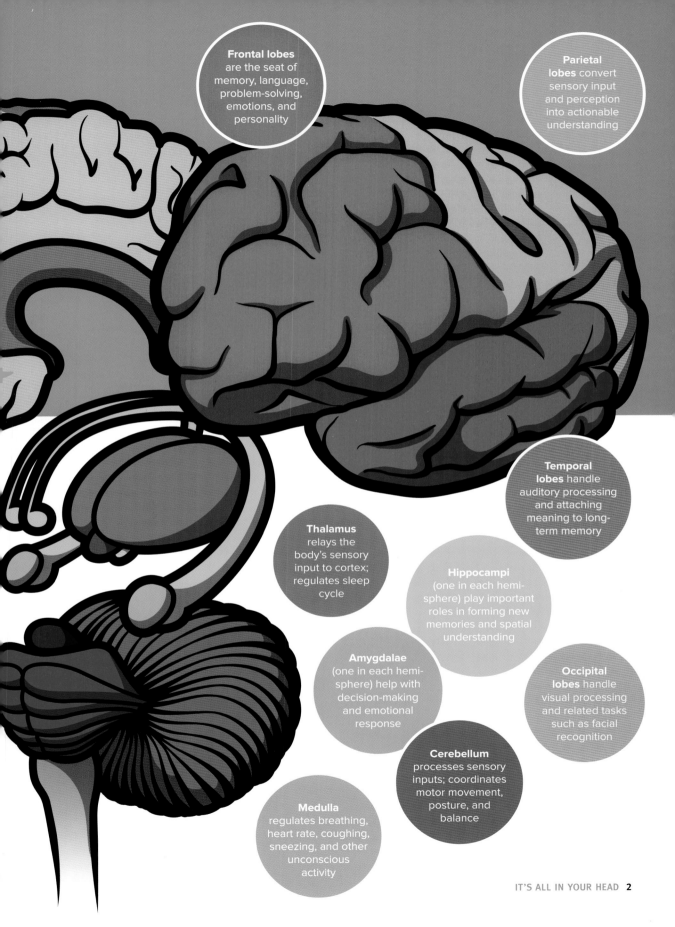

Frontal lobes are the seat of memory, language, problem-solving, emotions, and personality

Parietal lobes convert sensory input and perception into actionable understanding

Temporal lobes handle auditory processing and attaching meaning to long-term memory

Thalamus relays the body's sensory input to cortex; regulates sleep cycle

Hippocampi (one in each hemisphere) play important roles in forming new memories and spatial understanding

Amygdalae (one in each hemisphere) help with decision-making and emotional response

Occipital lobes handle visual processing and related tasks such as facial recognition

Cerebellum processes sensory inputs; coordinates motor movement, posture, and balance

Medulla regulates breathing, heart rate, coughing, sneezing, and other unconscious activity

CHAPTER 1

THE AMAZING BRAIN: WHAT IT WAS, IS, AND IS BECOMING

Who are you? Specifically, what are the boundaries—where does "you" stop and the outside world begin? It's not as simple as it sounds.

You might think of yourself as your contiguous set of cells, for example, but it turns out your body contains about as many bacterial cells (i.e., other life forms) as human cells. They're much smaller, sure, but there are tens of trillions of them: Do you really want to include them in your definition of you? Worse, even the cells that are yours are constantly being replaced. Defined as the sum of the cells you have now, only about 2% of "you" was a part of your body just a year ago.

But you're clearly "you." And the thing that's really unique about you, of course, is your mind.

This endlessly fascinating machine—and its squishy, compact carrying case, the brain—holds a whole lot of what's important about you: your memories and preferences, your skills and beliefs, your dreams and desires, your history of wisdom and folly. Millions upon millions of thoughts and impressions and feelings and opinions have piled up over years, packed in with incredible efficiency, yet are still changing materially every second of every day. If you lose your arms or legs or lose the use of your body below the neck, it's a terrible loss of valuable tools and

MIND BENDERS

5 *Minutes without oxygen before a brain begins to die*

15 *Watts of power generated by your awake brain*

7,000 *Number of human brains stored in Harvard's "Brain Bank"*

Your brain generates just about enough power to keep an actual light bulb lit over your head.

capabilities...but you're still you. If your mind begins malfunctioning, though, people might start saying things like "Grandma? Well...she's not really here anymore."

So let's start our exploration of the mind with the brain itself: the electrically charged bag of meat in your head. If we're going to have any hope of understanding how our minds work—how ideas spring up, how memories evoke emotions, how dreams inspire action, how to fix what's broken and celebrate what's merely unusual, how to hack our own potential and capabilities for pleasure and productivity—we have to start with the wetware.

Let's peek under the hood and take a look at your brain.

Your brain weighs just under three pounds, if you're a healthy adult. (Although, if we're weighing your brain, you may not be the perfect picture of health.) It's about 60% fat and wrinkled like a walnut. Fed by thousands of miles of blood vessels, it sucks up 20% of all the energy your body takes in and boasts at least a hundred trillion electrical connections, a truly unthinkable number, equal to all the stars in a thousand Milky Way galaxies. The varying strengths and patterns and reinforcements of those connections are what represent everything you've ever known and seen and said and dreamed and imagined.

Is There a Difference Between the Right- and Left-Brain Hemispheres?

There are differences between the two hemispheres, but they're not nearly as profound as conventional wisdom would have you believe. It's true that your left lobe is typically more involved in language and math, and that your right side does more duty relating to spatial navigation and visuals. But both sides participate in most tasks, especially complex ones, and the idea that people more generally predisposed to logic and analytics are "left-brained" and creative dreamers are "right-brained" is lazy gibberish. A 2013 University of Utah study of more than 1,000 brains looking for just this effect found exactly nobody preferentially using one side or the other.

As a matter of fact, in some transcendently awful cases, people have had an entire hemisphere of their brain removed due to an accident or radical surgery and lived to function quite normally. The astonishing degree to

which the brain is capable of dramatic rewiring is known as **neuroplasticity**. In fact, "rewiring," as a convenient metaphor, understates the power of what's going on. There's no real equivalent to wires in your **cerebral cortex**, only pathways and patterns embedded in the tissue of your brain—pathways that can and do change, for any number of reasons.

Whatever else you are, you are almost infinitely flexible.

The right and left hemispheres of your brain work in harmony to produce the smooth, graceful activities we characterize with being human. Say you want to murder a mosquito on your knee. The activity itself is directed by your **motor cortex**, the part of your brain that controls muscle movement. (Reach up and tap the very top of your head with both hands, then slide the fingertips down toward your ears. Beneath the hair and scalp and skull, you just outlined the motor cortex; the right motor cortex controls the left side of your body, and vice versa.) But the motor cortex is only part of this team effort. You first need the goal defined, by the **frontal lobe**, and information on your hand's current position, provided from the **parietal lobe**, and remembered details on how swatting works, from the **temporal lobe**, and coordination by the **cerebellum**, a sort of air traffic controller making sure your hand slaps that mosquito, rather than, say, knocking over your coffee mug.

"Left-brained" and "right-brained" individuals may be a myth, but left- and right-handed people are real, and their brains demonstrate intriguing, if not thoroughly understood, differences. Lefties—about 10% of the population—on average have more brain symmetry and a larger **corpus callosum** connecting their lobes, which may lead to better communication between the right and left sides. This can work to their advantage: They tend to recover from **strokes** faster than righties, for example, as their more evenly-distributed brains are better at adapting to the new reality. But they're also more susceptible to disorders like dyslexia possibly because their language processing is more distributed across the brain, which turns out to be less efficient. Right-handers' brains seem to develop along a more standard path in utero, while lefties' brains stray from the dominant developmental path. Later, as adults, lefties are more likely to be gifted mathematicians and architects, but suffer more than their

A CLOSER LOOK AT YOUR **DELICIOUS BRAIN**

Glial cells and neurons up close

The cerebral cortex forms the bulk of your brain: two parallel hemispheres, each traditionally divided into four sections, front to back: the frontal (this handles movement, smell, speech, and higher-order functions like judgment and communication), the parietal (taste and touch, including your self-awareness of your body), the temporal (hearing, emotion, memory), and the occipital (sight). Aside from hosting the great bulk of your active thoughts and passive memories, the cortex hosts various higher-order functions like self-awareness and language.

Nestled between your cerebral hemispheres are some important specialized giblets, including the hippocampus (involved in memory and navigation), the pituitary gland (a "master gland" that releases hormones that control other glands), the olfactory bulb (smell), and the basal ganglia (an ancient tangle that controls motor functions, such as street-dancing and parade waves).

Tucked in at the base of your skull, underneath the cerebral cortex, the cerebellum coordinates muscle movements. Lower still in the brain stem, and closer to your body, the medulla oblongata takes on the thankless job of maintaining sub-conscious activities like breathing, heart rate, and sleep, as well as involuntary actions like sneezing and coughing.

As this image demonstrates, the left and right cerebral hemispheres have long been romanticized as odd-couple skullmates managing different functions.

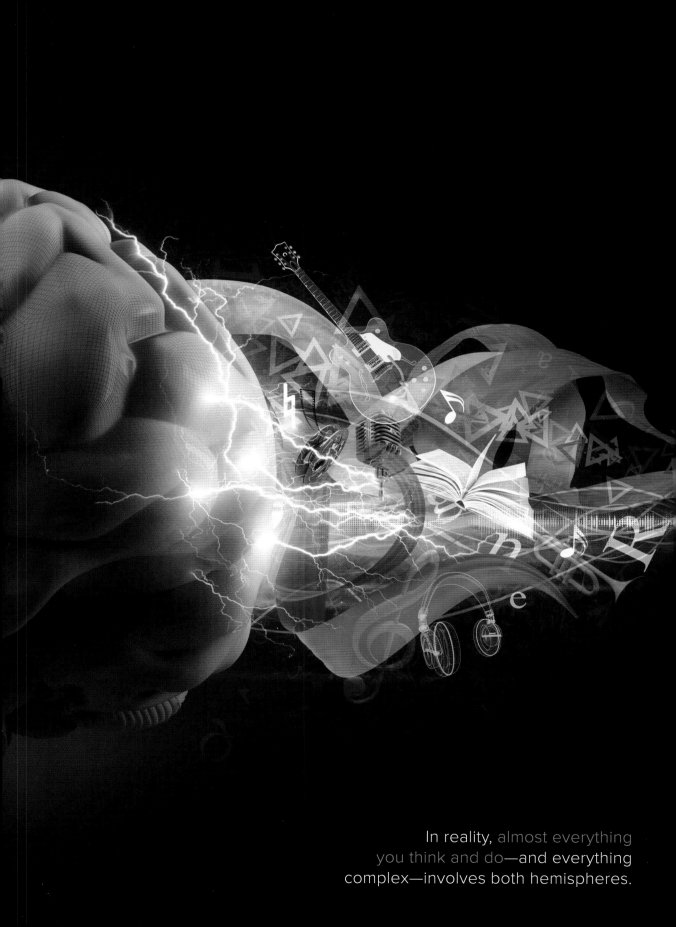

In reality, almost everything
you think and do—and everything
complex—involves both hemispheres.

share of allergies, migraine headaches, and autoimmune disorders. Righties, on the other hand, are less likely to demonstrate cognitive disabilities or behavioral problems, and earn 10-12% more income annually than lefties.

Why is Your Brain Wrinkled?

Your wrinkled brain is evolution's elegant solution for maximizing the surface area of your brain without having to build you a gigantic skull. Most of your meaningful brain activity—sensory processing, speech, and decision-making, for example—actually takes place in the outer surface of your brain, called the cerebral cortex, and that famously wrinkly surface packs a lot of processing power into a relatively small volume. This is your **gray matter**, and the **white matter**—nerve fibers that connect gray matter to other parts of gray matter— sits beneath it. Think of your cerebral cortex as a crumpled, rubbery sheet about the thickness of two stacked pennies: If you could

unfold it all the way, it would be about the size of a small pillowcase (and you'd have one hell of a party trick).

This folding happens as brain growth outpaces skull growth in the womb; at birth an infant's brain wrinkliness is about the same as mom's. Lower-order animals like rats tend to have smoother brains; at the other extreme, those brainy dolphins have even more convoluted cortices than humans. That said, intelligence is a complicated affair, and more brain wrinkles don't necessarily make one human smarter than another.

The visual similarities between the wrinkles of the brain and of the walnut did not go unnoticed by the credulous ancients. According to the generally hokey theory, the Doctrine of Signatures, foods and herbs that looked like parts of the body offered special benefits to those parts; therefore, walnuts must be good brain medicine. But coincidentally, that one actually checks out: Walnuts are today consid-

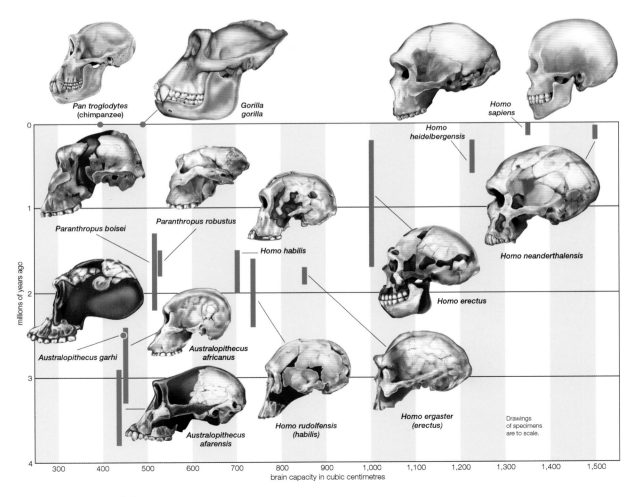

Pan troglodytes (chimpanzee)

Gorilla gorilla

Homo sapiens

Homo heidelbergensis

Paranthropus boisei

Paranthropus robustus

Homo habilis

Homo neanderthalensis

Australopithecus garhi

Australopithecus africanus

Homo erectus

Australopithecus afarensis

Homo rudolfensis (habilis)

Homo ergaster (erectus)

Drawings of specimens are to scale.

millions of years ago

brain capacity in cubic centimetres

ered high-value brain food that can improve memory and may help counteract age-related cognitive decline like that produced by **Alzheimer's disease**.

Are Bigger Brains Smarter Brains?

Yes and no. Primate brains are twice as large as you might expect for mammals of our size, and, sure enough, every chess grandmaster in history so far has been a primate. From an evolutionary perspective, these giant heads of ours are a liability: Human childbirth is painful and dangerous in the wild, and human babies are helpless and dependent for quite a long time, compared to, say, fawns. Yet for all of that, bigger brains conveyed enough of a survival advantage to take humans to the top of the food chain. Our brain has tripled in size over the last 5 million years or so—from Australopithecus through *Homos habilis, erectus*, and *sapiens*—corresponding with a gradual improvement in intelligence. Civilization and big heads may have developed in tandem, since reliable help from other nonthreatening humans made childbirth safer, which helped even more big-headed kids survive in each generation, and so on.

So far so good: In humans/**hominids** over the long haul, growing brains have historically correlated with growing intelligence. But the human brain is actually shrinking now, believe it or not. Our hat size seems to have peaked in the Stone Age, and since then we've lost on average about a tennis ball's worth of volume. Among the possible reasons for this shrinkage is the notion that civilization itself has made it less necessary to be super-smart to survive, which has allowed the not-so-bright to overstay their welcome in the gene pool. ("Thanks for calling tech support. Can you start by verifying that your computer is plugged in?")

Interestingly, the male brain weighs about four ounces more on average than the female brain: the equivalent of two Snickers bars. And yet there's no difference in average IQ between the genders, no matter what your drunk grandpa claims. It turns out a lot of the storied Mars/Venus differences between the male and female brains, such as the canard that men are better at math than women, appear to be based not on innate abilities but on history and socio-

BIG BRAINS...BAD LUCK

Although Neanderthal brains surpassed ours in volume, most of their capacity was devoted to body control and vision, leaving less mental real estate available for complex cognition and social processes.

cultural expectations, such as keeping math-based fields closed to women for a long time. One experiment demonstrated that men are more likely to think they scored higher than they did on math tests, while women are more likely to think they scored lower; another showed that women perform better on math tests when using a pseudonym.

The bottom line on size: The surface area of your brain—not its size—is a better predictor of actual relative intelligence.

Are Humans the Smartest Animals?

Unless toads or butterflies are seriously slow-playing their hand, it seems humans are indeed at the head of the class. There are larger brains than yours in the animal kingdom, for sure, but none that can begin to compete with you on Trivia Night. What seems to matter more than brain size, in terms of intelligence, is relative brain-to-body size, as well as total **neurons** and other details of the brain's internal structure. For example, neurons, the brain's building-block nerve cells, are typically found in higher numbers in larger brains within any one family (like hominids, or great apes—you are smarter than your distant orangutan cousin). But when you compare across families (say, humans vs. dolphins), the different brain structures complicate the equation. Elephant brains are three times the size of ours, and they have three times the number of total neurons. Unfortunately,

SIX ANIMALS THAT ARE **SMARTER THAN** YOUR LITTLE BROTHER

It ain't the size of the skull—it's how you use the creamy filling.

We great apes may be top bananas when it comes to intelligence, but the delta between us and the "lower animals" may not be as dramatic as we'd like to believe. Here are a few contenders for Earth's second-smartest fauna.

CHIMPANZEES, our closest evolutionary relatives, are very social, if violent, creatures: They make spears, use tools, cooperate, take care of their impaired, and mourn their dead. They can even outperform humans at certain cognitive tasks like short-term memory tests.

HONEYBEES have brains the size of a pinhead, yet can count, read symbols, and communicate complex flight vectors to the hive. Connoisseurs of color, bees can differentiate between Monet's or Picasso's paintings, and may have rudimentary self-awareness.

PIGS are remarkably clever when it comes to finding food, using tools to locate food, and other food-finding tasks. They can recognize themselves in mirrors (most animals think it's a different animal) and have been taught to play videogames and adjust a thermostat.

DOLPHINS are capable of communication, playfulness, and other attributes we normally consider human, and have a high neuron density like humans and a cerebral cortex even more convoluted than our own. Parading this animal's intelligence ultimately backfired for Sea World.

DOGS may not be smart enough to differentiate between "food" and "poop," but they've grown very good at understanding humans. Rushing to a spot you are pointing at in order to find a missing ball isn't trivial—it's something other animals, by and large, can't do successfully.

CROWS can solve problems that would perplex certain humans we know. In one of Aesop's fables, a thirsty crow can't reach the water at the bottom of a jug, but drops stones in until the water level rises high enough to reach; experimenters replicated this "fable" with actual crows—and it worked.

How Do Thoughts Work?

Okay, buckle up—we've got to define some hardware first. As referenced in the *Your Noodle In A Nutshell* infographic on page 1, the four key elements of your neurological wiring are neurons, **axons**, **dendrites**, and **synapses**. Neurons are like little trees, with multiple outstretched branches called dendrites and a single long trunk called an axon. Axons can be microscopically small, but some overachievers are many feet long, like those of the sciatic nerve that connects your spinal cord to your big toe. The axon communicates information out to other neurons, whose dendrites receive that communication. The junctions between the axon of one neuron and the dendrites of another are called synapses, and it's across these tiny gaps that communication, at its most basic level, takes place.

Axons are insulated, though not continuously so, with a fatty bubble wrap called **myelin** that helps these millions of "cords" pack tightly together without short-circuiting and conduct electrical signals at a suitably brisk pace. Without the magic of myelin, it's been estimated that your spinal cord would have to be as thick around as a tree trunk to work. Myelin damage can lead to crossed-signal issues; sufferers of multiple sclerosis (MS), in which the myelin deteriorates, exhibit symptoms such as blindness in one eye, numbness or tingling, and difficulty with coordination.

The last pieces of the puzzle are **neurotransmitters**—the chemicals that communicate between neurons by jumping the razor-thin 30-nanometer **synaptic cleft** between them. When a neuron "fires," the axon releases specific chemicals that spray across this teeny no-man's-land gap and are picked up by specialized receptors on the next neuron's dendrites. These chemicals are called neurotransmitters, and they do a range of exciting things.

Those are neurons, and their sequential firing, in new or remembered patterns, ultimately gives rise to conscious thought. Neurons are reasonably blank slates that specialize according to your inputs—and with some 86

only 3% of them are located in their cortex, where what we call intelligence lives.

Brains generally seem to have evolved from earlier, simple structures like **nerve nets**, which merely propagate a sensation ("ouch!" "yum"), and help brainless animals like jellyfish respond to their environment. In vertebrates, an evolutionary process called **cephalization** centralized and strengthened these neural clusters into actual brains, conveying all kinds of survival advantages. But the animal kingdom is vast, and evolution has granted some animals altogether different ways to organize their intelligence. Around half of the neurons of an octopus, for example, are clustered in its eight arms. Imagine what consciousness would be like, distributed across your limbs like that. When an octopus loses an arm, the severed limb will coil and uncoil for hours, possibly feeling its own peculiar kind of disembodied pain.

billion neurons in the human brain, they can get incredibly specific: Researchers found a "Jennifer Aniston neuron," for example, that fired when a test subject was shown any picture of her, and a "Sydney Opera House neuron," and so on. The neuron patterns are so specific, in fact, that researchers recently have made strides reading people's minds simply by watching these super-specified neurons and inferring, when they fire, what the subject is thinking about. ("The subject seems to be imagining Jennifer Aniston heading into the Sydney Opera House...")

Thought happens when sequences of firing neurons register as memories, new sensations, or creative ideas. With each individual neuron capable of connecting to tens of thousands of others, your brain has 100 trillion connections, an unthinkably vast network. If you're still imagining your brain as a supercomputer, you're severely underselling it. Measured in terms of computational capability, it's 30 times faster than the world's most powerful supercomputer today—yet stuffed into a much smaller package. If you could find a way to rent out the processing power of your head, it's been estimated you could easily earn $100,000 a day.

No wonder the brain's been called "the most complicated object in the universe."

And let's not forget, while we're wallowing in the magic, that our machine, unlike your MacBook Pro, self-assembles in the womb. The **ectoderm** begins to take shape as a proto-brain and spinal cord, and development takes off from there. Neurons develop

Fight or Flight? You sweat bullets when confronted by your evil boss because your autonomic nervous system that deals with stressful situations evolved to prep you for physical battle or danger evasion.

> "*You don't fear the bear when you see it— you start running, and THEN you feel the fear.*
>
> –**William James**, American philosopher & psychologist

at a furious pace—250,000 every minute, at times—and migrate into the correct parts of the brain as the fetus grows. As the cerebral cortex outpaces the inner white matter in growth, it begins to crumple. The baby's brain contains virtually all the neurons it will ever have, though not all the connections, at about 40 weeks. To get all this through the birth canal, the skull famously has to finish its assembly after the squeeze through, like a miniature ship waiting to unfold its sails until it passes through the neck of a bottle.

It's all quite miraculous.

What's the Difference Between "Instinct" and "Reflex"?

"Instinctual" behavior has a little more decision-making to it than automatic or **reflexive behavior**. Breathing, for example, is a reflex—it's automatic and driven by your **brain stem**—it's **unconscious** unless you choose to focus on it. Instinctual behavior, on the other hand, is

reflex and instinct. It can be a reflexive activity (when you're being tickled, and literally can't control it) or instinctual (when you're laughing at a good joke and aren't making a conscious choice to laugh but can rein it in when the boss walks in). Both versions activate similar regions in the brain, with two notable exceptions: Humorous laughter activates the **nucleus accumbens**, associated with our pleasure center, while tickling instead activates the **hypothalamus**, our **fight-or-flight** regulator. Somehow, tickling is tied to the survival instinct. It's been surmised that tickling may have evolved as a way to teach our children to defend their softest spots, and may have been an evolutionary "building block" for the kind of laughter we actually enjoy today. Laughter, in small bits, is woven curiously deeply into the fabric of everyday conversation, where it's believed to serve a variety of unconscious social purposes, like bonding. But it remains instinctual, not conscious: You can't make yourself laugh, and people rarely laugh alone.

> I used to think that the brain was **the most wonderful organ** in my body...

conscious action you take for reasons you don't quite understand. The brain is a busy place, and top-level attention, or consciousness, can't be given to everything; through consciousness, our cleverly organized brain is just giving us the topline. But there's a world of processing going on beneath the surface. Instinct lets you act, consciously, before you have all the facts. Reflex lets you act unconsciously, before you have any facts. Instinct tells you to get out of the elevator when the vaguely creepy guy gets in. Reflex pulls your hand off the hot stove before you know you need to.

Laughter can illustrate the difference between

What is the Fight-or-Flight Response?

When something dreadful and surprising warns us of danger, our body unconsciously awakens the **autonomic nervous system**—an ancient mechanism for instantly rallying the mind and body for quick, decisive action. The emotions-and-decisions center, **amygdalae** (there are two, one in each hemisphere), conspires with the chemical-weapons-factory hypothalamus, activating the pituitary and **adrenal glands** to release **hormones** that shut down nonessential processes (like the immune system and digestion) in favor of increasing blood pressure and glucose production. Other

...then I realized who was telling me this.

Emo Philips

hormones, including **adrenaline**, course through your system, prepping your mind for focus (you lose peripheral vision, similar to tunnel vision, and some hearing) and your body for action (your heart pounds and breath quickens; you sweat and get goose bumps). **Cortisol** makes stored energy available to muscles, increasing muscle tension for effective fighting or fleeing, and makes chemicals available that help repair tissue in anticipation of injury.

What we remember is the fear. Interestingly, though, "fear" seems to be an invention of the conscious mind to understand your body's strange instinctual reaction; it tends to arise only after your brain and body do a **subconscious** analysis of the situation. You don't fear the bear the instant you see it—your body starts running (or at least preparing itself to run), and after you notice your body's reflexive response and start trying to make sense of it, that's when you feel the fear.

It's all very stressful over the long haul, and adrenaline is toxic if you get too much. You can literally be scared to death, because strong emotional responses can lead to heart attacks: This is why people die during sex or even after a hole-in-one. There was an increase in cardiac deaths in New York in the weeks following 9/11.

How all that neuro-machinery and chemistry come together to produce you—your specific, individual collection of thoughts and dreams and opinions and beliefs and self—is a complex matter indeed, one of the most profound questions we can ask. Here's what we know so far: Sensory input starts in the womb—by the time they're born, babies already recognize their mother's voice and are attracted by the smell of breast milk—as does reflex action, like a newborn's "startle" reflex that leads him to fling his arms out in fear and then contract into a hug. (This is why we swaddle them into tight bundles, by the way.)

Motor-controlled crawling starts at around nine months; speaking at around a year; self-awareness at around 15-24 months. (How do we know? It's the moment when a bit of rouge smeared on a baby's nose in front of a mirror will lead them to reach for their own nose rather than the one in the mirror.) First sorting of shapes and colors occurs around age 2, first understanding of letters and numbers around age 3, first recognition

FOUR WAYS WE CAN **PEEK INSIDE** YOUR SKULL

From ancient times we've wondered what part of the brain does what—and we've started to get very good at it, thanks to a whole raft of brain imagery tools. Here are four neuroimaging tools and what they show us.

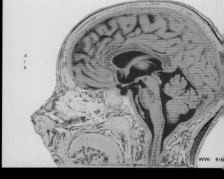

▲ **fMRI (Functional Magnetic Resonance Imaging) monitors brain activity** due to changes in blood oxygenation (active brain areas consume more oxygen, increasing blood flow to those areas). It is useful for mapping which brain regions are activated in the course of various activities.

◀ **EEG (Electroencephalography)** monitors the brain's electrical activity (a.k.a. "neural oscillations") through electrodes on, and occasionally in, the scalp. Its high temporal resolution (down to the millisecond) is good for diagnosing epilepsy, sleep disorders, and knowing whether or not someone is in a coma.

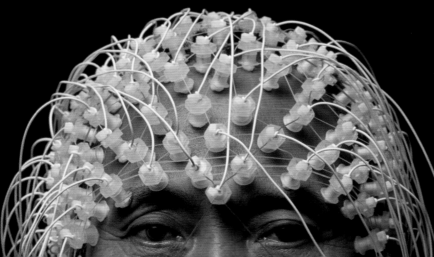

your parents are irredeemable idiots around age 13. It was assumed for a long time that the brain stopped developing in adolescence, but now we know the brain continues development into your late 40s. If the bulk of the people you meet on Tinder seem immature, this may be a reason.

And all along the way, in and around the developmental milestones, the child is consciously and unconsciously building something that might loosely be called a self. It happens to all of us: You gradually but continuously build a vast repository of specific memories and preferences; you form and test opinions and beliefs; you master a bewildering array of skills ranging from general threat evaluation to on-the-fly pig Latin translation to the ability to perfectly snatch a single tissue out of the box without ripping the tissue or lifting the box off the back of the toilet. Each new sensation or idea is weighed against your history and preferences, further specializing your unique experience. When you and I watch the same movie, eat the same meal, or hear the same conversation...we don't. Not really.

The miracle is on: You become distinctly, and quintessentially, and forever, you.

One intriguing outcome of all of this is that every new thought shapes the mind that holds it. Your mind evolves and changes, adapting to everything you feed it, every minute of every day. It receives inputs through the senses, and compares them with what it knows, and decides what's important, and suppresses the rest (how long has that car alarm been going off?). This produces a picture of the world that dramatically cuts down on needless detail, but provides enough important information so that you can function in the world and make informed decisions.

And somehow, out of this constant self-reflective processing of infinite data points arises something we call consciousness. You know that you exist, and you know that you know; this is possibly unique in the animal kingdom. It may seem miraculous to the observer; it is proof of God to some. But Marvin Minsky charted long ago in his Society of Mind—and computer programs have borne out many times—that our minds' jaw-dropping complexity can arise through simple rules and iteration. Magical though it may seem, neurons and axons and

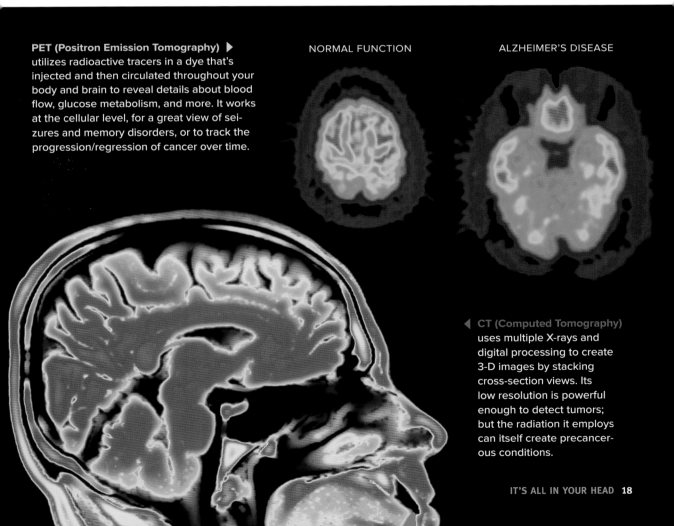

PET (Positron Emission Tomography) ▶ utilizes radioactive tracers in a dye that's injected and then circulated throughout your body and brain to reveal details about blood flow, glucose metabolism, and more. It works at the cellular level, for a great view of seizures and memory disorders, or to track the progression/regression of cancer over time.

NORMAL FUNCTION

ALZHEIMER'S DISEASE

◀ **CT (Computed Tomography)** uses multiple X-rays and digital processing to create 3-D images by stacking cross-section views. Its low resolution is powerful enough to detect tumors; but the radiation it employs can itself create precancerous conditions.

SCREEN **TEST**

Did Limitless *get the percentage of our brains we use right?*

The sci-fi thriller *Limitless* explored the conventional wisdom that we only use 10% (although the movie claims 20%) of our brains, but it simply isn't true—every part of the brain has a known function. But the brain is amazingly plastic, and people can live normal lives even when an entire lobe of the brain is removed.

dendrites in their millions, firing and squirting neurotransmitters, really can—and do—lead to movie scripts and breakthrough inventions and Beyoncé Super Bowl dance numbers.

Welcome to the power of the mind.

In the next chapter, we're going to tackle sensory input—how your mind, sitting alone in the quiet dark of your skull, knows what it knows. (We'll also look at some of the fun ways it can be fooled.) But first, there's just one more important question to address about the brain itself.

Why Are Brains So Damn Tasty to Zombies?

Ironically, in *Night of the Living Dead*, George Romero's black-and-white masterpiece and the undisputed father of all zombie movies, no brains are eaten. The concept came into the canon later, in the sequels. Dan O'Bannon, the director of the sequel *The Return of the Living Dead*, suggests that perhaps it eases their pain—brain matter is, after all, high in soothing **serotonin**. (See *Your Noodle In A Nutshell* infographic on page 1.) Maybe the concept is terrifying for us because it's the final taboo. In an extreme Donner-party type situation, you might be convinced to see a stray arm or leg as meat...but a brain? To ingest someone else's personality and history and all that? It would be like eating their soul.

Unfortunately for Earth's edgiest gourmands, eating brains in the real world can cause a host of issues—as we know from the records of cannibals who ritually did it. Among the perils are **prions**, tiny protein particles that can transmit brain-targeting pathogens including **mad cow disease** and, in the case of human brains, a laughing sickness called **kuru** that can make you laugh literally until you die. What crazy thoughts roll through your head in those long moments of spontaneous madness, as you laugh so hard you're choking to death?

Recommendation: Try the salmon instead.

CHAPTER 2

SENSORY INPUT, ILLUSION, AND MAGIC: HOW WE KNOW WHAT WE THINK WE KNOW

Your brain is a fantastic analytic and computational engine. But on its own, it's blind, deaf, and dumb. It's like a dizzyingly complex, polished, well-oiled piece of machinery sitting by itself on a table in the dark.

The brain itself doesn't "know" anything real; its connection to the outside world is through sensory inputs, like the big five (sight, sound, smell, taste, and touch), together with a lot of secondary inputs that make your stomach rumble, raise the hairs on the back of your neck, let you know your leaned-back chair is about to tip over, and so on. Taken together, your sensory inputs provide a nonstop flood of data for your brain to process and hammer into a coherent world you can understand.

The *lingua franca* of the brain, as we've seen, is made up of electrical impulses and chemicals. In a very real sense, it doesn't matter whether the inputs come from your eyes, your ears, or the

MIND BENDERS

14 — *Number of miles away you can detect candlelight*

5.2 — *Millions of times you blink in a year*

6 — *Weeks you can expect to enjoy that "new car smell"*

You can't taste without your saliva. Don't believe it? Dry your tongue with a paper towel, then eat a dry snack like a cracker.

er") is a condition where sensations from one sensory input are experienced as if they come from another. For those affected, sounds can have smells, shapes can have tastes, and other curious but beautiful oddities. "Synesthesia is a love story between the senses," explains professor and avant-garde artist Hugo Heyrman, who maintains a web page to connect synesthetes and their experiences with curious muggles like us. Synesthesia, in various forms, has powered the perceptions of a number of prominent artists, including Billy Joel, Mary J. Blige, Kanye West, Marilyn Monroe, and Vincent Van Gogh.

Vladimir Nabokov, author of *Lolita*, was a "grapheme-color" synesthete, meaning that different letters—and the sounds they made—had distinct colors in his head. "The long 'A' of the English alphabet," he said grandly, "has for me the tint of weathered wood." Interestingly, his wife was affected with synesthesia too, but saw letters in a different palette. And their son had yet a third set of letter-colors that seemed

edge of your hand on the hot stove—whichever sense first picks up the stimulus converts it to a distinct electrochemical signal that shapes the initial sensation. (Luckily, your nose will tell you that the milk has soured before your tongue will.) Distinguishing smells from sounds, sights, tastes and so on helps us organize our impressions into understanding. But to really get a handle on how it works, let's first take a look at a strange border condition called **synesthesia**, where the lines aren't so clear.

Synesthesia (Greek for "to perceive togeth-

Synesthesia is more common than people think…about 4% of us have some form of it.

in some cases to be a blend of both: Vladimir saw "M" as pink, his wife saw it as blue, and their son purple.

But that's not the crazy part: Researchers discovered a few years ago that *we are all born with synesthesia*. When you are visiting your aunt's newborn baby in the hospital, and you're watching the baby checking out the room and the visitors, understand that the baby is literally incapable of sorting out whether her mother's voice is a smell or a sound. That adorable confused look really means something like: "Well, *this* is generally pleasant. I wonder what the hell's happening."

The trick is growing out of it.

Research is still ongoing, but adult synesthesia may be a result of "cross-wiring" between adjacent areas of the brain—the areas that process colors and shapes are tangential. About 1 in 2,000 people have synesthesia, and it can be passed from parent to child (as with the Nabokovs). If you're curious what the experience is like, there are free apps that try to mimic the effect through filters, and an online how-to on creating a "synesthesia mask" that lets you smell colors; hypnosis can reportedly produce the effect, too.

THE REDEYE EFFECT IN PHOTOS IS THE LIGHT FROM THE FLASH **ILLUMINATING THE BLOODY BACK WALL OF YOUR EYE**

Sense #1: Sight

Sight is far and away our predominant sense; the extraordinarily complex processing that vision requires takes up about 30% of your entire cortex. (Hearing, in contrast, takes less than 4%.) Your two optic nerves each have about a million fibers, streaming electrical signals out the back of your eyeballs like parallel transatlantic cables.

What we call the "visible spectrum" is only a human limitation; light has many, many wavelengths, and humans see only a narrow range between what we call red and violet. (Remember the "Roy G. Biv" of the rainbow: red, orange, yellow, green, blue, indigo, violet?)

Below red on the spectrum is infrared light—the waves your TV remote uses to communicate with the TV. You can't see infrared with your eyes, but you can feel it; you recognize it as heat. (Some snakes, like boas and pythons, can see heat, multiplying their overall creepiness.)

Above violet on the spectrum is ultraviolet (UV) light. Imagine being able to see the rays of the sun: Ultraviolet light is responsible for freckles and sunburn, and it's what makes your rave paint glow. Bees, butterflies, and some other animals can see in ultraviolet, and we recently discovered a hidden world of natural art, where flowers have developed ultraviolet "landing strips," invisible to you and me, that guide potential pollinators.

HOW DOES THE EYE WORK?

Here's how the eye captures data to feed your brain. Light reflects off of something—a kitten riding a Roomba, say—and that complex re-

Carrots don't improve your eyesight. It was a myth perpetuated by the British, during WWII, who who didn't want the Germans to know they had radar.

Sensory perception begins in the womb, with wee fetal synesthetes smelling/tasting mommy's food and hearing/feeling her voice. For the non-synesthetes among us, development often leads to clear differentiation among what we've come to call the five major senses: sight, sound, smell, taste, and touch—five unique systems for converting analog data from the real world into electrochemical signals for processing. They're each fascinating, and each contributes to the brain's overall understanding of the world around us.

WHAT BLIND PEOPLE SEE

(Hint: it isn't just blackness.)

You can't really explain "blue" to someone who was born blind. But for people who were born sighted and went blind, or—much rarer—born blind and became able to see, it's possible for one person to describe both states. So: What's it like to be blind? What do you "see" in your head? Turns out it's different from one blind person to another, based on degree or type of blindness. Some see blackness, some see brightness and colors that can be beautiful or distracting or even intrusive. And as for those born blind, who have no experience from which to draw, it's like asking you what a magnetic field looks like.

HOW TO FIND YOUR BLIND SPOT

flected image enters through your clear cornea, then through your pupil, the tiny black hole in your colored iris, which dials open and closed to adjust the light coming in. The image then hits your lens, which focuses the rays (and inverts the image) onto the back of the eye, called the retina. This sheet of nerve tissue contains millions of rods and cones, and each interprets its piece of the visual signal and converts it to an electrical signal. Cones are clustered more toward the center, and handle color and fine details; rods are distributed more around the outside, and handle peripheral vision, motion, and dim lighting situations. All of their data gets piped out the back of each eye, through your optic nerve, and ultimately to the **occipital lobes** at the back of your head.

But that massive continual feed of visual input is just the beginning of sight, because knowing what you're looking at is complicated. You evaluate brightness and color, shape and clarity, the size of objects and their motion relative to you, whether they're new or familiar, foreground or background, and so on. The science on this is not yet settled, but there seem to be a number of separate systems for processing this information in parallel. Strokes and other localized injuries can have disastrous effects on one specific part of vision, rendering somebody with otherwise "normal" vision suddenly colorblind or unable to recognize faces.

Sight has been described as a system of sensory reasoning in which all of this separate visual processing comes together in real time to paint a coherent picture of what's going on out there. When different pathways present conflicting or missing information (remember, this is all happening in milliseconds), the brain compensates by making educated guesses on the fly—filling in the gaps so you don't stand there in confusion while the puma leaps. Essentially, everything you see is illusion—a story your brain is telling you about the world.

SECRET You see a black-eyed susan in traditional yellow and black. But flowers like this have evolved secret ultraviolet target colors that can only be seen by helpful pollinators like bees and butterflies.

Every human has a blind spot, in each eye, where you have no rods or cones to process light. To prove it to yourself, close your right eye and stare at the blue dot. Now slowly move your face closer to the page, continuing to stare at the blue dot until the red dot disappears from your peripheral vision. At that point, the image is focused right on your left eye's optic nerve, where there are no cones or rods—and that's why you can't see it.

It's a system designed for speed, not accuracy, and it can be fun to expose some of those gaps and short-cuts your brain is making and watch your mind work its magic on you. Here are just a few examples:

Your mind expects light to come from above. Evolution has hardwired your brain to interpret all light as sunlight. That's why you look so creepy when shining a flashlight from beneath your chin while sharing your spookiest ghost story—the unusual lighting angle unnerves your listeners' brains.

Your mind creates depth when it can't find it. Your brain combines your eyes' 2-D inputs into a 3-D "reality," which is the secret behind many illusions and much art—it's easy to draw a cube or other object that looks 3-D because your brain naturally supplies the depth.

Your mind expects motion. Stare at a still object long enough and it will start to appear flat and two-dimensional. This is because you're hardwired to detect motion and change; when it isn't happening, your mind stops looking for it.

Your mind "fills in the gaps" with a kind of logic. Touch your index fingers together in front of you and stare past them at this page; you should see a third finger appear like a hot dog between your index fingers. You're really seeing double-images of the ends of your fingers, but your mind looks for the kind of distinct objects that populate the real world.

Color blindness is a deficiency of the color-sensitive cones in your retinas. Normal eyes have three kinds of cones that respond (roughly) to wavelengths we recognize as red, blue, and green. If your "red" cones are damaged, you see a more sepia-like world, where something the others call purple might look blue to you. Color blindness affects around 8% of American men but less than half a percent of women (it's carried on the X chromosome, so women have two chances to avoid it). There's no cure, but there are now sunglasses available that can help many affected see and smartphone apps that offer help as well.

Cats and dogs are color blind—they do see color, contrary to popular belief, but their cones are less sensitive than ours. Same with bulls; they charge the red cape because of the motion, not the color. At the other extreme are birds, with amazing color appreciation (which you might infer from the world-class plumage of peacocks and parrots). Birds also have incredibly sharp vision, by and large—a hawk can spot a mouse in motion from hundreds of feet in the air. We humans have worked hard to expand the limits of our visual capabilities, conquering the near and far with microscopes and telescopes, developing infrared goggles and ultraviolet light readers.

Sense #2: Sound

Music, birdsong, ocean waves, the wind in the trees, traffic and sirens, the laughter of children, and the neighbors bickering again. It's a noisy world out there. And our brain tries hard to sift through only what's important.

Sound waves are different than light waves, however. Sound waves are pressure waves, a regular pattern of alternating high and low pressure signals. They don't carry molecules with them like the wind does—they're more like regular patterns of disruption, like ripples in a lake. And like ripples, sound waves from different sources can interfere with one another constructively or destructively, bringing the complexity of a symphony or a busy restaurant to those waiting jughandles on the sides of your head.

HOW DOES THE EAR WORK?

The outer ear is nothing more than an amplifier: Cup your ear with one hand to double its powers, then imagine how much quieter things would sound without it. The real magic happens in your inner ear. The

When you look toward the sun with closed eyes, ever see little worm-like "floaters" in your vision? These are remnants of the hyaloid artery that fed your growing baby lens and then, having no way back out of your completed eyeball, dissolved inside it as best it could.

THINK ABOUT THIS

*Nothing we use or hear or touch
can be expressed in words that equal
what is given by the senses.*

Hannah Arendt

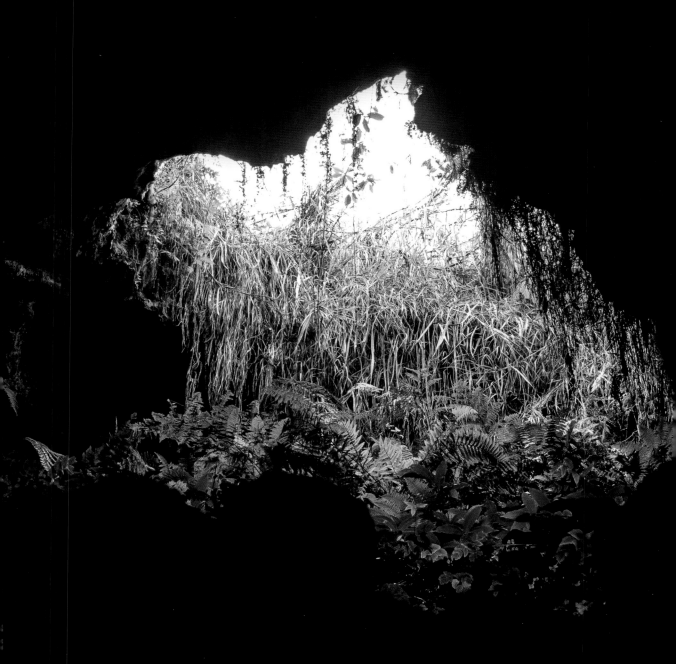

incoming sound wave channels through your ear canal and vibrates the tympanic membrane at the end, which you know as your eardrum, that thing you're strongly encouraged not to pierce with a Q-tip. On the other side of the eardrum is the middle ear, where sits a trio of tiny bones, the smallest in the human body, nicknamed hammer, anvil, and stirrup for their shapes. (Hammer's the one with the baggy pants.) These vibrate sympathetically with the eardrum, recreating its sound waves in a pressure-controlled environment so as to enter the cochlea, a.k.a. the inner ear, a snail-shaped container filled with fluid and lined with some 25,000 nerve endings that translate the pressure-based audio waves into electrical impulses that fire off to the brain, via the **auditory nerve**, for processing.

Two other things worth mentioning inside your ear: First, the Eustachian tube connects your ear canal with the back of your nose, so you can equalize pressure on both sides of your eardrum for hi-fi sound. When pressure is uneven, your ears feel clogged, like they do when you're about to land in an airplane. Second, your inner ear also contains an organ that helps you maintain balance, called the bony labyrinth: When the balance information here doesn't accord with what your vision and other sensory inputs are telling the brain, it can result in dizziness, nausea, and, in some cases, vertigo. According to one theory, roller coaster nausea occurs because your body assumes

you're hallucinating from poison—evolution has conditioned our bodies to react to such sensations by throwing up.

IS IT TRUE THERE'S A HIGH-PITCHED WHISTLE THAT ONLY DOGS CAN HEAR?

Yes, and many other animals have super-hearing, too. Sound is measured in hertz (Hz), or cycles per second, with low Hz corresponding to low-pitched sounds like your roommate's diabolical snoring, and high Hz to the high-pitched shriek he makes when you shake him awake. Humans can pick up sounds roughly in the 20 Hz-20,000 Hz range—a piano would need 5 more keys on the low end and 15 more keys on the high end to catch the whole range. But elephants and moles can hear lower sounds than this (cleverly called "infrasounds"), and, yes, cats and dogs can hear higher-pitched "ultrasounds" up to around 40,000 Hz. Dolphins and bats supersize this ability, hearing well over 100,000 Hz. And congratulations, Millennials: You can hear a higher frequency than your older counterparts—the high end of our audible range starts to degrade after the age of 25. Enjoy it while you can.

The way that sound waves move affects our hearing. At low frequencies, sound waves are long—up to many feet long—which is why church organ pipes are so massive. At the high end of human hearing, they're only about as long as your thumb is wide. Unimpeded, sound waves can travel far—picture yourself in the nosebleeds at a Genesis concert—but almost

anything can impede them, like the thin plastic on the outside of your headphones. Sound waves reflect easily, too, giving us echoes; arts theaters are designed specifically with multiple 3-D surfaces that return the stage's sound to you in appealing ways. Curved interior spaces such as that outside New York's Grand Central Oyster Bar and Restaurant are known as "whispering galleries" because sound waves bounce along the wall and can carry a whisper to people far along the curve; in a perfectly elliptical room, sound created at either focal point will bounce off the curves of the wall and return to the other, and you can eavesdrop on people at one focus by standing at the other. (Just keep your voice down—it works both ways.)

HOW DO NOISE-CANCELLING HEADPHONES WORK?

When two sound waves meet, they can reinforce constructively or destructively. When it's constructive—that is, when the peaks meet the peaks and

Your ears are not the same!
Your left ear is better at music and your right ear better at understanding language, because of where those processing centers reside in your brain.

the valleys meet the valleys—sound waves get louder. But if you can induce them to reinforce destructively—make the peaks meet the valleys—they cancel one another out. Active noise-cancelling headphones hide a tiny microphone in the outer shell that picks up the ambient noise and cleverly produces a waveform that's the polar opposite and destructively interferes with it. Result: What you hear from the outside world is close to silence, leaving you to focus on your beloved Radiohead.

Sense #3: Smell

To see and hear, your senses pick up wave activity. But with smell and taste, we move into a trickier world where solids, liquids, and gases are making direct contact with your body. Did you ever wonder as a child how smells go up your nose? When you smell something, are tiny pieces of that thing actually disengaging and entering your nose? It seems pleasant enough, when considering cupcakes for example, that tiny bits of cupcake were atomized and being scooped out of the air by your nose. Of course, there are a lot of things that give off smells, and there are many whose molecules you really don't want in your nose.

And turns out, for better or for worse, that's pretty much what smelling is all about.

HOW DOES THE NOSE WORK?
In humans like you, millions of sensory receptors are arrayed in the mucus at the back of your nasal cavity, poised to catch any smellable molecules that might sweep in. These olfactory receptor cells come in 400 or so different types (dogs have closer to 1,000), each of which can only bind with certain aroma molecules. These highly specialized cells have tiny hairs called cilia to help receive these molecules and extend long axons out the other side, through the bone of your skull, directly into a highly organized processing unit. When a receptor detects the molecule it's right for, nerve endings in the receptors alert the brain and it's go time.

But you can smell much more than 400 things, of course—your college dorm room alone probably had more than 400 noxious odors in it. It's been estimated, though the science isn't settled, that you may be able to distinguish among around a trillion different odors. The system is a little loosey-goosey: One sensory receptor type can bind to multiple different smells, and one smell can activate different types of receptors, and different signals are sent depending on how strong or weak the bond is between molecule and receptor, or how close the smell is to a related smell (which is how a glass of wine can have hints of cinnamon and blackberries despite having no such ingredients).

These signals can produce a tremendously nuanced and detailed picture, but only after some serious processing, so you and many other mammals have evolved a specialized organ for this. Called the **olfactory bulb**, it lives conveniently behind your nose, and it takes in

the axons from the nasal cavity, amplifies even low-concentration smells that affect just a few receptor cells, and sends these processed signals to the **olfactory cortex**. Geography is everything, because before these smells can get to the brain they have to pass through a dicey neighborhood. The olfactory bulb is part of the **limbic system**—an ancient area of the brain connected to memory and emotion. And that's where things get interesting.

IS IT TRUE THAT SMELL IS THE CLOSEST SENSE TO MEMORY?

In a sense, this limbic system, which includes the amygdala, the hypothalamus, the **hippocampus**, and other goodies, processes emotional reaction and memory formation. Because smells route through here, they can immediately "take you back" in a visceral way you can't quite understand, and before you can even identify the odor. Remembering a song or a person's face is a more cortex-driv-

en proposition; smelling is a deep-rooted thing that pre-dates human sentience. That pizza smell does reach the cortex eventually, of course, to be put together with other information: Now you might recognize the smell of garlic and pepperoni, and infer that it's probably hot (based on the strength of the smell), and other bits of human cleverness. But for a brief moment there, you were enjoying a pure animal sensation.

Our sense of smell, sadly, is not tops in the animal kingdom; there may well be other animals living in your house that can smell better than you can. But it's still insanely impressive and can deliver profoundly detailed information to your brain at an unconscious level. The scent of women who are ovulating can raise men's testosterone levels, and in one fascinating experiment, gay men were reportedly found to prefer the sweat of other gay men to that of straight men. In another, women asked to evaluate

GO FOR BAROQUE
The aesthetic excesses of old theaters were sometimes about dispersing sound waves. But recent research says shoebox-shaped concert halls sound best, because our ears are on the sides of our head—perfect for receiving waves reflected directly off long flat walls.

men purely by smelling their sweaty T-shirts preferred men whose DNA was very different from theirs and therefore were more likely to produce children with strong immune systems.

The sense of smell is duller in cognitively impaired people, like those with Down syndrome, and loss of smell is an indicator of, and under consideration as a test for, early onset Alzheimer's disease.

Did you ever think it was strange that sharks have nostrils, since they're clearly not breathing underwater? Smell is even more mission-critical to animals in the wild, and for many animals, like sharks, smell is the predominant sense. It sounds like an urban myth, but some sharks can smell their prey's blood and bodily fluids at a one part per billion concentration, the equivalent of one drop in an Olympic-size swimming pool. And with their nostrils on the sides of their snout, they can even get

smell: The simple act of smelling food before you eat it can heighten your abilities, and focusing on therapeutic aromas you find pleasant (say, coffee and donuts) multiple times a day can heighten the response of receptors to those smells, according to a University of Oklahoma study. Experimental subjects exposed to a single floral scent for more than two minutes improved their ability to distinguish it among multiple flowers.

Sense #4: Taste

Let's talk about poison.

There are quite a lot of different things in the world that a curious human might decide to put in his mouth, some of which provide nutrition, and some of which can kill you dead. As omnivores who eat everything we can, it's especially important for us to be able to distinguish between what's helpful and what's harm-

The reason food tastes bland when you have a cold is because **smell is a big part of taste**, and the extra mucus in your nose keeps the food molecules from reaching the olfactory receptors at the back of your nasal cavity.

directional signals, like you do with your ears, and triangulate the struggling, doomed fish. (Imagine being able to do *that* with your nose.)

WHY PREGNANT WOMEN HAVE AN IMPROVED SENSE OF SMELL

It's true that pregnant women are temporarily gifted with an extra-sharp sense of smell, for better or worse, during the nine-month hormone bath. You could be banished from the room if you enter with a strong-smelling thing like a spoonful of peanut butter. The effect reportedly has something to do with the extra estrogen produced during pregnancy (the effect is also seen in non-pregnant women with heightened estrogen levels), but beyond that science doesn't quite know why it happens. It's also notable that women born with no sense of smell—an uncommon condition called anosmia—are less likely to experience morning sickness when pregnant.

HOW TO IMPROVE YOUR SENSE OF SMELL

It's relatively easy to improve your sense of

ful. Evolution's answer was to reward us for the helpful and punish us for the harmful. You know this reward/punishment system as taste.

Humans can generally distinguish five families of flavors: sweet, salty, sour, bitter, and umami, which is a kind of savory, meaty taste. We used to think each taste owned a particular zone of the tongue, but that turned out not to be so. Taste is complex, and is itself complicated by other factors, notably smell but also texture, temperature, and other factors. You're naturally attracted to sweet and savory things, because those tastes in the wild usually mean sugar and protein, a.k.a. power. Similarly you're naturally averse to bitter and sour foods, which in nature are often toxic or rotten.

As for salty? Interestingly, your desire for, or aversion to, salty foods at any time depends on your body's pH balance—i.e., whether you need it or not. Sometimes you feel like a nut; sometimes you don't.

While some believe other types of flavors, like

THE SMELL TEST
Smell is so important to taste that people with anosmia reportedly can't tell vanilla and chocolate ice cream apart.

Your sense of smell is at its best in spring and summer, because the air's moister, and early in life; it begins to decline while you're still a teenager.

How To Improve Your Sense of Smell

It's actually relatively easy to improve your sense of smell: experimental subjects exposed to a single floral scent for more than two minutes improved their ability to distinguish among multiple flowers.

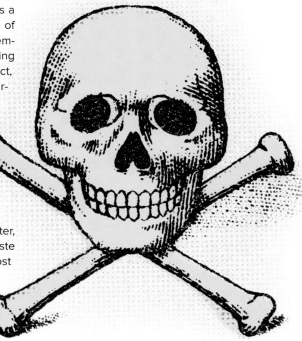

Bitter tastes, in nature, often signify things that are poisonous to humans...untold millions died to bequeath you an aversion to bitter. Enjoy your hoppy IPA!

"fatty" and "spicy" and "metallic" and "watery" are worthy of consideration, the big five predominate. What you think of as a complex taste, like scotch or crème brûlée, is a combination of these basic inputs, plus mitigating smells and textures and temperatures and memories and so on. The magic of experiencing this as a distinct "taste" happens entirely in your brain. In fact, you can easily fool your tongue: There's a West African "miracle berry," for example, which coats your tongue and temporarily changes your mouth's receptors so that vinegar tastes like apple juice. Scientists can now induce the ability in mice to taste specific flavors through direct brain stimulation; within a generation there'll probably be a smartphone app that can make any food taste like whatever you want it to.

Smell and taste emerge early in the synesthete fetus. Later, smell and taste take a backseat to sight and sound. And taste declines over time: By the time you're elderly you may have lost half your taste receptors or more, and you might find yourself adding a lot more salt and spices, or having less interest in food, which is one reason old people have trouble keeping weight on. The elderly also lose nerve endings and mucus production in the nose, which compromises both their senses of smell and taste.

HOW THE TONGUE WORKS

Think of your tongue as a meaty probe, designed specifically to reach out, like a curious snake, and test your food for poison and nutrients, before you commit to swallowing it. As you chew, bits of food start to dissolve in your enzyme-rich saliva, and little bumps on your tongue called papillae begin to rise up, creating tiny wells for saliva so the taste buds at their base can detect and analyze the molecules. The sides of your tongue, and the front and back, are more taste-sensitive than the middle. You have about 2,000–8,000 taste buds in total, and

they only live for around 10 days before being replaced, which is why burning your tongue on hot soup isn't a lifelong disabling event.

About half of your tongue's sensory nerves are multipurpose, and respond to multiple tastes, in a preferential order—so one nerve might be most sensitive to salty, then slightly less so to bitter, and even less to sweet. Impressions from these cells give us a flavor's complexity. The other half of your tongue's nerves respond to only one taste (say, bitter), and impressions from these give us a flavor's intensity.

When sensory nerve cells in a taste bud are activated, they release chemical messengers that alert cranial nerve cells, which fire off a message to the base of your brain—the **medulla oblongata**, responsible for

moderating unconscious activity like heart rate and digestion. (Like smell, taste is connected to the autonomic nervous system, which is why bad food can make you gag.) Here, that message splits in two on the way to consciousness, with some signals going straight to the sensory perception part of the cortex, joining with impressions from smell, and others first merging with signals from the "touch" department (like temperature and pain). What you experience as the taste of pizza is the combination of all these inputs into one general impression: salty and savory and maybe a little sweet, familiar and—ow!—hot!

Besides the taste buds in your mouth, you have taste receptors in the back of your throat and a little ways down the esophagus (you may be conscious of those when you gargle or throw up), and in the back of your nasal cavity. Infants have them on the roof of their mouth as well.

HOW DO FLIES TASTE?
Flies taste with their feet, because absolutely everything about flies has to be disgusting, apparently.

HOW DO SHRIMP TASTE?
Delicious!

Sense #5: Touch

The soft caress of a lover, the unforgiving grit of sandpaper, the fuzziness of a kitten's belly, the searing of your lap from the spilled hot coffee. Your skin is famously your body's biggest organ, and keeping a taut boundary between yourself and the outside world, with all its dangers and delights, is a busy job. Touch is complicated in many ways. It isn't centered on one sensory organ (or a handy matched pair); it's all over your body. It measures many different kinds of sensation, including pressure, temperature, pain, itchiness, stickiness, and more.

The whole complex of nerves comprising touch is known as the **somatosensory system**, and it includes other important aspects as well, like knowing when your stomach is full and how your limbs are positioned (which the brain learns from nerves in your elbows and knees). Unlike all your convenient head-based sensory organs, touch is active everywhere, so this is where the spinal cord—a network of nerve fibers that make sure the rest of your body can communicate sensory impressions to the head—comes into play.

There are many kinds of highly specialized receptors in and under your skin that contribute touch information, from super-sensitive hair follicle receptors (you can blow on your arm to feel them at work) to free nerve endings that sense changes in temperature to **nociceptors** that send a pain sensation when they encounter damage. There are millions and millions of these receptors mapped all over your body, and they give you amazing precision; your fingertips can feel a bump on a smooth surface, for example, of just one 25,000th of an inch.

All of these touch sensations reach your brain's cortex, where a fairly unexpected and miraculous thing happens: they are mapped to a tiny version of you, called the **cortical homuncu-**

> "*I know this steak doesn't exist.*
>
> *I know that when I put it in my mouth, the Matrix is telling my brain that it is juicy and delicious. After nine years, you know what I realize? Ignorance is bliss.*"
>
> —Cypher, *The Matrix*

Your fingers get pruney in water because it increases their surface area and improves their ability to handle wet things.

lus. (Look at the drawing on the opposite page, if you dare; it's quite terrifying.) This map lets you correlate an incoming sensation like pain or pressure to an exact location on your body. You might note two interesting features: First, the amount of surface area each feature gets here is a function of its sensitivity, not its size: The butt, large but fairly numb, is barely represented here, while supersensitive lips and hands get much more real estate. Second, while many features are contiguous (the foot is near the leg, and so forth), there are a few things out of place: the genitals are down by the feet, and the hand is right next to the face. Nobody yet knows why.

Touch is about much more than just understanding the boundaries of the world around us. Touch between humans, particularly, is critical socially and developmentally. For in-fants and toddlers, frequent parental touching and holding increases positive emotions and creates more emotionally stable adults; it even helps kids grow physically. We apparently need mechanosensory stimulation; those deprived of it tend to grow up antisocial and reserved and are at increased risk of a host of additional ailments. Studies of orphaned infants (who often don't get as much touch) have shown that just 10 minutes of touch a day significantly reduced regurgitation. Even as adults, we need touch; it's been shown to boost the immune system of those suffering from colds, for example.

Pain in the Brain

Pain is your body's warning system, designed to keep damage to a minimum. It's a distressing sensation you feel when the **anterior cingulate cortex**, **thalamus**, and insula respond

Every year, from age 18 on, **we lose about 1% of our sense of touch.**

We're good at reading clues from social touching. In one study, participants could correctly guess the emotional state of a stranger touching their arm from behind a curtain.

to harmful stimuli, both physical and emotional.

It's possible in some situations to control pain through mental effort, or condition yourself to feel pain less: Positive thoughts and **meditation** work for some, and situations like battle or excitement can temporarily mask pain. At the other extreme, sufferers with a condition called CIP, for "congenital insensitivity to pain," are unable to feel pain at all; those with the condition can often injure themselves in normally unthinkable ways, like biting off the tips of their tongues.

The flipside to pain is **itch**: It's your body's way of urging you to scratch or swat away a potential irritant, like a bug. When your body picks up on a light touch or tickle, nerve cells far from the site of the irritation release a polypeptide that triggers the impulse to itch. Itching and pain are related in that they both involve serotonin; however, while serotonin reduces pain, it increases itchiness. This explains why scratching an itch feels good initially, but then feels progressively worse.

WHAT IS PHANTOM LIMB PAIN?

Phantom limb pain occurs when a limb is missing (after an accident, say), but your cortical homunculus is still receiving signals as if it were there, tricking your conscious mind into feeling sensation from where the limb used to be. The strange condition isn't as rare as you might think—it affects an astonishing 80% of amputees. And they aren't making it up or imagining things—**fMRI** scans show brain activity similar to that of people with intact limbs. One amputee, who claimed she could actually see a milky-white, transparent missing limb where her arm and hand had been, agreed to try scratching her cheek with the missing hand while in an fMRI machine. The results demonstrated not only activity in the motor and visual sections of her brain (i.e., as if her hand had moved and she saw it moving), but also a sensory response from the cheek, as if it had been scratched.

How does phantom limb pain work? Some believe the area formerly served by a now missing limb is "hungry" for sensation, and that

some nerves adapt and invade adjacent zones of the cortical homunculus. The homunculus maps the face next to the hand, for example, so to test this theory, one researcher tried touching the face of phantom limb pain sufferers, and some claimed to feel the sensation in the missing limb, not in their face. In another clever experiment, doctors set up mirrors that recreated a missing limb by reflecting the intact limb in a mirror—so an amputee missing a right arm would "see" it in a mirror reflecting their left arm—and phantom limb sufferers reported pain reduction.

Our sense of touch has all kinds of subconscious influence over our decision-making. In experiment after experiment, researchers have found they can easily manipulate people's feelings and perceptions based on nothing more than adjusting what the subjects are touching. Heavier objects can feel more important to us than objects of lighter weight, for example, so employers may be more likely to consider seriously a résumé printed on heavier stock versus lightweight paper. Additionally, people handling rough objects (vs. smooth ones) are more like-

TRY IT You can cure an itchy throat by scratching the insides of your ears.

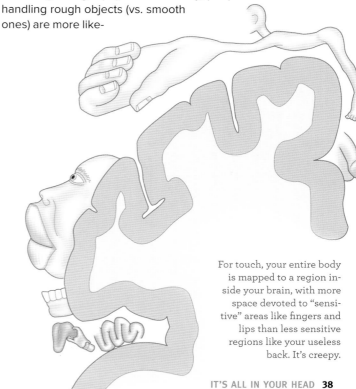

For touch, your entire body is mapped to a region inside your brain, with more space devoted to "sensitive" areas like fingers and lips than less sensitive regions like your useless back. It's creepy.

Free samples at retailers like Costco are so effective that they've been found to **boost sales as much as 2,000%.**

ly to see neutral social situations in a bad light, and are more likely to infer that other people are in a bad mood. In one experiment, subjects on their way to evaluate a neutral job candidate were interrupted by a staged accident and temporarily asked to hold a participant's mug that held either a hot or cold beverage. When the evaluations came back, researchers found the cold mug holders had rated the candidate as more cold and distant than did the hot mug holders. (Many of these "priming" experiments have been disputed, it's worth noting.)

Party Trick: Confuse Your Motor Cortex

Rotate your right hand and your right foot in a clockwise motion. Now try switching just the hand to counterclockwise, keeping the foot moving clockwise. You'll probably find the foot switching to follow the hand. Why is this so hard? Because your motor cortex sends a general signal called a "turning curve" that tries to direct all the muscles on one side of the body. Try moving your right hand and left foot in opposite directions instead; you may find it's easier.

Super-Sensory Perception

That's an overview of the five senses, and most of us are endowed with roughly similar versions of them. But occasionally, someone will rise up claiming to be endowed with more than their fair share of sensory abilities. Take sommeliers and wine reviewers, who can supposedly taste insanely subtle distinctions among wines. Or musicians like Stevie Wonder who reportedly have "perfect pitch," and claim to be able to sing and/or recognize any note without using a reference tone. Are these superpowers real? And if so, can they be learned?

Take "perfect pitch," or as it's called today, "absolute pitch," the ability to accurately name and/or reproduce a note that's played or sung. It's estimated that 1 in 10,000 have this ability, and display subtle differences in brain structure relative to people without it. But "perfect" maybe a misnomer; in one fascinating experiment, researchers demonstrated that by slowly changing the notes in a song while subjects with absolute pitch listened on, they could easily manipulate their sense of pitch, so that the listeners mistakenly thought a song was still in

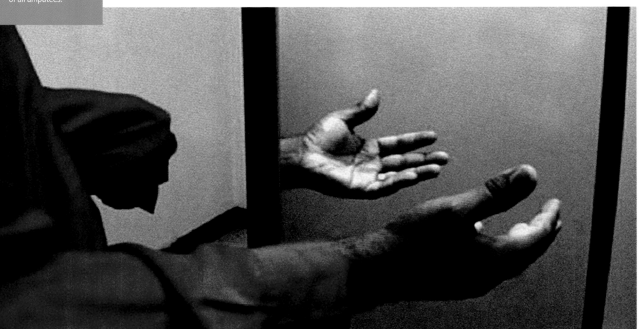

MIRROR THERAPY
"Phantom limb pain" may seem like a bizarre border condition—but it affects around 80% of all amputees.

What did you see first?

tune when it was significantly flat or sharp.

As far as wine experts go, it's absolutely possible to tell the difference between good wine and great wine, but the exactitude implied by the 100 point ranking system is not valid—researchers were able to get critics' scores of the same wine to vary by several points based on whether the wine was presented in a nice or plain bottle, for example.

For another ability involving physical differences, consider tetrachromats. They're gifted with four kinds of cones in their eyes instead of the factory-issued three; as a result, while *you* can only see a million colors, *they* can theoretically see 100 million (and would be a nightmare at the paint store). Imagine dividing a wall into squares painted with extremely subtle variations of white: A functional tetrachromat would see a multicolored checkerboard, whereas you and I would see a uniformly white wall. The catch is the functional part: While as many as 12% of women (it mainly affects women) have the fourth set of cones, they are virtually all nonfunctioning, and have normal vision. As with the similarly semi-mythical supertasters, tetrachromats are rare or nonexistent; none who claim the power have ever been confirmed by peer-reviewed testing.

Sensory Fatigue

Sensory receptors of all types can get fatigued, leading to a temporarily diminished signal. A persistent sound, like traffic outside the window, can become background noise you forget about as you get used to it. You might leave the kitchen where you've been cooking all morning to get the mail and only notice the delicious smell upon your return. The color **photoreceptors** in your eyes can get temporarily "burned out" after a matter of seconds, but they can just as quickly refresh themselves.

How Do Birds Find Their Way Over Thousands of Miles?

The animal kingdom contains all kinds of sensory superpowers. Take whiskers or echolocation: Each has its own magic. Whiskers don't help cats and dogs smell—they're tactile, registering fine vibrational detail from objects, air currents, and the like. Bats can echolocate, using sound waves from their mouths and noses to navigate in the dark and help them find food. It's worth noting that bats are not actually blind, and that humans, believe it or not,

The Strange Story of the Homunculus

A homunculus is a "little man," a medieval idea that each sperm cell contained a fully formed tiny person in it, who would grow into a baby when mixed with an egg. Of course, a male homunculus would have its own tiny sperm cells, which would have to have even tinier homunculi inside them, and so on, and so on. By medieval logic, this handily explained why we all have original sin: Adam had all of humanity literally inside him. Including you.

WHAT ARE AFTERIMAGES?

Stare at one color long enough and you'll fatigue the photoreceptors for that color. Then when you look at a white surface, those receptors will have weakened relative to the others, so you'll see the image in the opposite colors. See for yourself: Stare at the tip of Che Guevara's nose at left for 30 seconds, then look away at a white space. What do you see?

can learn to echolocate, too. Migratory birds' ability to find their way hundreds or thousands of miles with no GPS is another thing. It's long been assumed they use the magnetic fields of the earth to align themselves for summer and winter migrations, and science seems to bear that out, as migratory birds have iron in their ears and beaks, presumably for this purpose.

Illusions

The brain is designed to work very quickly to make decisions about lots and lots of data. It succeeds in this superhuman endeavor by cutting corners (like reducing the relative value of things in your peripheral vision), filling in gaps (like the blind spots in your vision), and making assumptions (e.g., the sound that's getting louder is probably also getting closer). Without these shortcuts, our good, reliable vision as we know it wouldn't be possible. But because they exist, we can use these and other flaws in the system to fool our brain into thinking it perceives something it doesn't. These we call illusions, if they're free—or, if we paid good money to be fooled, magic.

3-D Drawings: The Original Illusion

Cave paintings are the very oldest representations of human culture that survive from our earliest days. And even these first human drawings represent illusions; in particular, the illusion of three dimensions displayed in two. Cavemen drew outlines of deer, rabbits, birds, mammoths, and other humans. Why does an outline work? It turns out that the borders between things are an important deciding factor in our sight processing. The edges of lakes and cliffs are important, obviously, so we evolved special neurons to pay attention to edges.

We live in a three-dimensional world, and our eyes and visual processing system have had millions of years to evolve toward making better and better sense of what we see. Your spaced-apart eyes, for example, can feed your brain the slightly different vector information that lets you construct a sense of depth. With just a little shadowing, it's easy to create a 2-D drawing that any child will recognize as a 3-D object; this is the original illusion. In a similar vein, those classic red/blue 3-D glasses work by feeding your two eyes slightly different versions of the same TV or film image, tricking your visual processing system into constructing false depth for you.

How Do "Magic Eye" illusions work?

Other kinds of illusions betray other biases of our systems. The "Magic Eye" illusion—called a stereogram—works by presenting a repeating pattern that tricks your eye into thinking certain corresponding points in the pattern are part of the same object seen from your eyes' two angles. As it would do with any real-world object, the mind gamely constructs a logical depth that corresponds to where this false object seems to be, which is on a plane parallel to but behind the page, and once it's focused there, the differences in the repeated pattern manifest as the hidden 3-D image.

The Dress That Broke The Internet

Is it black and blue, or white and gold? In February of 2015, a dress caused major buzz across social media—and in the scientific

To adjust your eyes quickly to a darkened room, keep one eye closed while you try to see with the other, then switch them...**the closed eye will have adjusted itself.**

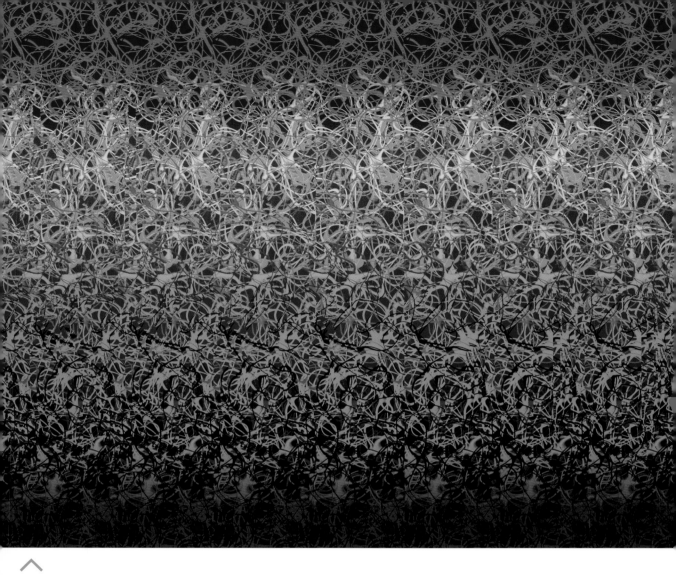

Hold the image up to the tip of your nose, focus on one point, and relax your eyes so the image becomes blurry. Slowly move the image away from your face, keeping your eyes relaxed and your focus on that point. If you just can't see the image no matter how hard you try, we've put the answer in the index.

munity as well. The statistics were fascinating: Older people and women were more likely to report seeing The Dress as white and gold, while younger people and men were more likely to declare it black and blue. One theory was that how you saw the dress was a function of whether you had most recently been in natural or artificial light, with those who were in artificially-lit environments more likely to see the dress as black and blue. Others pointed out the special case of the color blue, as there's gathering evidence that this is the color we're most likely to misperceive especially in settings of blue-tinted light like fluorescents. Bottom line: We still don't know, but it recalls the longstanding philosophical conundrum that there's no way to know definitively that you and I are seeing the same colors. In some cases, it seems, we definitely aren't.

Audio Illusions

Not all illusions are visual, though those are the easiest to showcase in books. One audio example is Shepard tones, an audio effect created by overlaying multiple sets of tones that seem to confound logic by rising forever in pitch—one set of sounds gradually rises in pitch but shrinks in volume, as a second set of tones gradually becomes audible, starting at a lower pitch and a quieter volume but eventually taking over, and so on, giving the illusion that the music keeps going up and up and up. Led Zeppelin's "Kashmir" exploits the effect; it was also featured in the movie *The Dark Knight* in the audio of Batman's motorcycle, so it could sound like it was continuously accelerating without changing gears.

Is There Any Evidence of Extrasensory Perception (ESP)?

Nope. Despite all kinds of testing, there's no evidence that there are people who can read each other's minds, know what people are going to do before they do, or any of that other mumbo-jumbo. The closest we have to mystery communication is menstrual-cycle syncing, which is not well understood but not exactly a valuable superpower. Animals may have **pheromone** communication, but humans will probably have to make do with Axe body spray.

But if you're willing to define "you" more broadly, you can start to extend your sensory abilities. This can be done through training (like teaching yourself to echolocate, for example), by equipment (like buying heat-vision goggles to see in the infrared spectrum), or by cyborgization (like implanting a magnet in your finger to be able to sense magnetic fields). Have fun in the TSA line with that one.

Our senses are a complex, interconnected system that gives our powerful brains something to go on. Organizing sensations that represent threats and opportunities from the outside world, monitoring body information from the inside, and putting it all together to present us with sensible "perceptions," the brain has a way to assess, instantly and more thoughtfully, almost any situation.

BE THE BAT You can actually train yourself to echolocate like a bat. Spanish researchers found that subjects taught to make a palate-click could tell when an object was in front of them or not after a few weeks, and could distinguish an item like a tree a few weeks later.

THINK ABOUT THIS

But it's still just the beginning: Unless you intend to re-experience absolutely everything every day, the brain has to store and recall what's happened today, so it can map new sensations to past perceptions, and begin to build a reasonable expectation of what will happen if you do *X* instead of *Y*. In short, you have to be able to *recognize* things—poisons and sexual opportunities, the feel of a blister, the taste of chocolate chip cookies. Recognition and memory are the keys to being able to make sense of new sensory inputs, and put them into useful categories for processing.

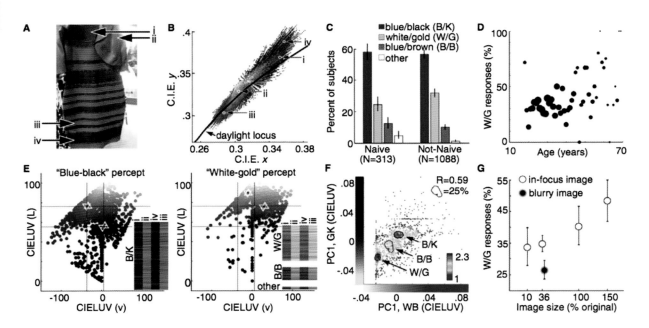

Anyone who ever wondered *"When we see green, are we both seeing the same color?"* found out the hard way when this dress took the Web by storm.

CHAPTER 3

MEMORY, RECOGNITION, AND TRYING TO FORGET

What is your earliest memory?

I remember the family Christmas of my early childhood like it was yesterday. My dad became a little kid himself every year: buying too many gifts, wrapping tiny things inside great big boxes for fun, leading my two sisters and me around the house with chains of paper as directional clues to get to something big.

There were tons of traditions, naturally: Stockings magically appeared at the ends of our beds overnight, to amuse us until it was legal to come downstairs. The same six-album set of old Christmas classics was played, in order, over the speakers. The room with the tree was walled off until after pancakes and coffee: no arguments, no matter how much we writhed in agony.

And then, Mom would lead us in (after Dad, so he could turn around and take the ritual photograph), padding in pajamas around the corner for that first glimpse of the magical tree wading in a glittering sea of presents. Our tree had colored bulbs, long gold braids of tinsel, and a gaudy frosting of silver garland right out of *Mad Men*. Mom's fudge perched, always, in a plate on the table in the living room for nourishment as we tore into the loot, one present at a time for maximum drama. I can see it all before me now, in the sepia tones of a fading Polaroid, the Instagram of its day.

But here's the thing.

Now that I've looked into this memory thing a little bit, I realize that I don't actually remember a single one of those Christmases. What I remember, instead, is all of them: a general sort of "Christmasness," an interlaced set of patterns that was rein-

MIND BENDERS

7,000

Maximum number of words the average person can remember

1 *Number of times W.A. Mozart reportedly needed to hear a song to memorize it*

2.5 *Petabytes of data your brain can store—about 300 years of video*

Children lose 40% of their early memories by age 7, and 60% by ages 8-9.

forced, year after year after year, by repetition of the same sensory experiences (Bing Crosby's "White Christmas," the taste of Mom's fudge, the silvery glitter on the tree), and by shared stories and photographs between Christmases.

It's likely the same for you. Human memory, it turns out, doesn't work anything like computer memory. An event—a song, an experience, a story—isn't stored intact and retrievable, like a digital song or video file. Rather, it's laid down as a foundational pattern on the first experience, a pattern that can be reinforced and associated with other patterns, that each has its own reinforcements, until the original is more of an inspiration. We don't "recall" events with the specificity that the word implies; it would be closer to say we remember remembering them. My memories of my old family Christmases are sepia-toned, because every time I looked at a Christmas Polaroid, it literally colored my memory.

Memory—the kind in living creatures—is an organic, chaotic process in which storage is easy, and retrieval sometimes a nightmare. "[It's] like a teenager's bedroom," says psychology professor Bennett Schwartz of Florida International University. "They know where everything is, but it's a complete mess." Nobody would design a computer this way. And yet this system conveys distinct advantages over computer memory,

starting with processing power: Your brain is about 30 times more powerful than the most powerful NASA supercomputer ever made.

Many of those little quirks about memory, from *déjà vu* to a song stuck in your head to trying to remember something that's "right on the tip of your tongue," aren't bugs in your system—they're features, and they're unavoidable because of the way human memory works.

But first, some groundwork.

What Exactly IS a Memory, Anyway? I Forget.

Sensory impressions and thoughts, as we know, are patterns of electrochemical activity: a set of particular neurons firing in series. A memory—the first time you rode a bike, say—is a collection of these patterns that can contain multiple elements: the wobble of the wheels, how old you were, what time of year it was, whether you felt liberated or limited.

Remember that each neuron can be connected to thousands of others. Some patterns involve many, many neurons; some neurons are involved in many, many patterns. And the patterns can overlap: Your memory of a particular Justin Timberlake concert and the lyrics of one of his songs, for example, can mutually reinforce and evoke each other.

The pattern a particular memory takes is chemically strengthened every time it's revisited. Neuron-to-neuron communication shuttles back and forth between the electrical and the chemical; a pattern revisited strengthens each connection along the way by marginally in-

Familiar scenes experienced year after year can fold distinct memories into one general memory.

Well, "never" is a long time. But elephants do have incredible memories—tops in the animal kingdom, except possibly for dolphins—and can remember a lot of practical (well, for them) information. These highly social creatures remember the locations of dozens of tribe members at one time, and where alternate food and water sources lie even if they haven't used them in decades. And while their eyesight isn't great, their smell-based recognition is keen, and they can easily distinguish, for example, between the urine of tribe members and outsiders. In one Tennessee elephant sanctuary, an elephant greeted a newcomer with incredibly tender reunion-like behavior—they checked each other's scars with their trunks, and so on. It turned out the two had briefly been in a circus together...*more than 20 years earlier.*

DO ELEPHANTS REALLY "NEVER FORGET"?

creasing the relative amplitude at the receiving end of each synaptic junction, making it more likely the signal will travel this way next time too. That's a memory.

The best way to remember a new experience—as you may well have discovered through trial and error—is to immediately revisit it on purpose, strengthening the original pattern. If you take notes during a meeting, notes you may never revisit again, that action nonetheless strengthens the baby patterns just laid down. If you repeat a story to someone right after the experience, or make a timely diary entry while it's still "fresh in your mind," you're reinforcing those initial impressions, improving the chances of the memory "sticking." (See Five Awesome Memory Hacks on page 60.)

And just as a pattern that's revisited is strengthened, a pattern that isn't is weakened. What you had for lunch on August 12th, 2008 is still in there somewhere, as a pattern, but odds are it's growing fainter and fainter as its component neural pathways are partly and gradually over-

written for more recent or more meaningful memories, getting more and more degraded and difficult to dig up. You *can* be reminded of something you've long "forgotten," as when an old friend finds just the right detail that brings an old anecdote back to life for you. But it becomes less likely as the years go by, as other patterns imprint themselves and get reinforced at the expense of earlier ones.

I recently came across *I Wish That I Had Duck Feet*, a favorite childhood book of mine that I hadn't seen for probably 35 years and had quite forgotten. It was originally written under a pseudonym, so it didn't make the general canon of Dr. Seuss books (*Green Eggs and Ham*, etc.) that you and I have seen more or less continuously over the years: in kids

Memory is a story you never stop telling yourself.

By Dr. Seuss*

*writing as **Theo. LeSieg***

Illustrated by B TOBEY

I Wish that I Had Duck Feet
50th Anniversary

DUCK FEET Even a long-lost memory, like a childhood book, can be recalled after decades if the "triggers" are specific enough.

rooms, in bookstores in passing, and ultimately in our own house, as our children grew. But *Duck Feet* totally took me back: I savored it, page by page, with memories flooding in. I would have said I'd completely forgotten it, but revisiting the book thoroughly rebuilt that pathway. Now I can talk about that book today, and connect it to my childhood as if there'd been no gap in that memory. It was brought back from the dead.

When you see your newborn child for the first time, you may hope to preserve the moment forever. But the information isn't "stored" in the sense of a "file," an exact set of data that goes into a drawer to be retrieved later. Rather, it's a sensory experience—seeing the face, hearing those baby noises—associated with emotions (love, protectiveness) and other signals of importance. Your memories aren't distinct phenomena from original experiences; rather, those original experiences, when recognized as important (in part because of associated emotions), are organized as a coherent, semi-retrievable story.

Ultimately a personal anecdote seems to be stored similarly to the way a pre-literate community stores its history. Each retelling strengthens the general story, but some details are deemed more important than others and are remembered better; "unimportant" details can be lost or contested over time. As with a community, unconscious biases can affect the way a memory's retold and remembered. As a result, that history—and your memory—can evolve over time, accidentally or on purpose.

How Do We Recognize Things?

When a pattern representing a sight, smell, or other sensory input from some real-world object or person has been encountered before, it's helpful to be able to pair that new perception with whatever happened last time. This is recognition—the first step toward memory. Without it, you'd still have taste to guide you toward strawberries and away from animal dung, for example, but you'd have to try each type of food over and over again, all your life; you'd never learn to avoid tigers or to trust friends.

Many neuroscientists divide recognition into two processes—others define them as two parts of the same process—familiarity and recollection. You're caught between the two every time you see a face you recognize but don't know why: It's familiar to you, but it's only later, when you can put a name to the face that you can truly say you recognize the person.

Recognizing individual faces, by the way, is a particular strength of humans, although sheep, elephants, and some other creatures have good facial recognition as well. It's estimated you can recognize around 10,000 different faces. We are social creatures, immersed in a chaotic world of trust and betrayal, friends and strangers, and aggression and compassion, and we tend to live in close quarters. Being able to recognize the face of a loved one—and, importantly, to be adept at quickly intuiting an emotional state from a facial expression—matters.

Interestingly, a normally functioning brain tries to assign just one identity to a face. In one experiment, an image altered to be 60% Marilyn Monroe and 40% Margaret Thatcher was interpreted as an "older Marilyn Monroe," and when the ratio was reversed the image was deemed a "sexier Margaret Thatcher." We also tend to recognize people from our own race better than those of other races, which bears implications for things like eyewitness testimony (where cross-race testimony should perhaps be even more suspect than same-race testimony).

How Do We Store Memories In the Brain?

Memory storage is often modeled as four phases. The first phase is **encoding**—when an experience or sensation is initially captured and temporarily laid down as a pattern in the hippocampus (you have two of these, one in each hemisphere). Second, that encoded proto-memory has to survive **consolidation**, a natural process that gets rid of unimportant things that fill our senses all day long, like the stuff you ignored on the supermarket shelves to the right and left of the Froot Loops you picked up and put in your cart. Third (sometimes combined with consolidation) comes **storage**, where the memory moves from the hippocampus to its permanent home in the cortex. And finally there's **recall**—when the memory is consciously drawn out to be re-experienced and strengthened anew.

PHASE 1: ENCODING

Encoding seems to happen something like this: Cursory impressions come in directly from the senses and are processed in their relative lobes (the temporal lobes process sounds, for example, and the occipital lobes process visuals). These separate streams are transferred to the hippocampus, where they are quickly woven together into a micro-experience. This "preattentive processing," named back when we thought it preceded conscious attention, registers as "sensory memory." From there the hippocampus helps us decide, in a reflexive sort of way, what's worth our attention, and hence, what will be shuttled on to the more familiar **short-term memory**, instead of discarded and overwritten by the very next thing you see or hear.

PHASE 2: CONSOLIDATION

Also known as working memory, short-term memory is conscious, and lasts a little longer—probably from fifteen seconds to a minute, though it seems to vary. This is where you temporarily keep the three errands you have to remember, or the phone number your roommate is calling out from the next room. You can extend the stay of an item in short-term memory by repeating the information to yourself, which

FACES IN THINGS In nature, a face offers uniquely important clues for threat and opportunity assessment, and evolution has equipped us to pay particular attention to them—to the point where we see them even where they're not, as the entertaining "Faces In Things" Twitter feed demonstrates. This is called "pareidolia."

neuroscientists call "rehearsal." But short-term memory is severely limited: When you give the bartender a drink order, or give the voter registrar your name and address, they remember what you said now, but in ten minutes they'll be unlikely to recall it. Psychologist George Miller determined that people can typically store seven or so items in their short-term memory simultaneously, though this can vary person to person. You can test your own limits with the short-term memory test on the opposite page.

The process of consolidation is how the brain decides which short-term memories are worth saving. The hippocampus, that sensory-processing workhorse, takes these inputs and evaluates what should and shouldn't be preserved in a couple of ways. It compares them to existing memories to see if you recognize them (if you knew the song playing at the car wash today you might remember it later; if it's an unfamiliar song you may not remember that music was playing at all). It reviews that input to see whether there was an emotional or physical component; if you bang your leg or get embarrassed, for instance, the memory gains importance and is more likely to survive the consolidation process. Proto-memories that pass muster are then sent on to storage.

PHASE 3: STORAGE

When short-term memory gets the right electrochemical signals that something's meaningful, personal, engages the body, or is otherwise

"important," it hands that proto-memory off to **long-term memory** for storage. Long-term memory is functionally infinite and can last your whole life. Everything you've done and known, as long as it's made it into long-term memory storage, is still in there somewhere. (Consider how, when rereading an old book, passages you'd long forgotten can seem completely familiar again.) But that doesn't mean it's accessible, and long-term memories can be difficult to retrieve, as you probably already know.

Long-term memory is stored all over the cortex and comes in many flavors. Different parts of your cortex are used for **visual memory**, **verbal memory**, and **spatial memory**. Other ways to categorize memories include **episodic memory**, which stores events ("I went to church today with my family"), and semantic memory, which stores facts ("Tequila is from Mexico"). **Procedural memory**, also known as "muscle memory," stores your ability to perform tasks and skills ("I can beat you at darts"). It's acquired through repeated behavior, and it's the reason you can play baseball or piano without stressing every time over the complexity of swinging a bat or playing a scale.

To the best of our understanding—which is still changing—new long-term memories seem to be forged in the hippocampus and amygdala, the emotional wrecks of your limbic system, then stored in the cortex. One patient whose hippocampus had been surgically removed

LOOK FAST Did you see the red and green strawberries at a glance, and not much else? That's your sensory memory at work: deciding what's most likely to be worth your attention (in this case, because it stands out from the background).

after an accident was unable to form new memories, though he retained some memories from before, which strongly suggests that a long-term memory is forged by the hippocampus working in concert with the cortex and, once the hippocampus version is overwritten, lives on only in the cortex. When participants in one study were asked to answer questions about the past 30 years of their lives, the hippocampus and amygdala were seen to be less active when they were retrieving older memories. Instead, there was activity all over the cortex—in the frontal, parietal, and lateral temporal lobes. One theory is that the hippocampus teaches the cortex a memory, then clears itself once the cortex is able to hold on to it.

PHASE 4: RECALL

Every time a memory is recalled it's relived, essentially. You know every guitar riff and cymbal crash of a song you love, because you've experienced it countless times, both actively (listening to a recording) and passively (remembering it in the shower), etching a deep and persistent pattern in your moldable mind. So to try to recall a memory physically is to try reliving the experience again. When we can't come up with a particular memory we know we should have, we instinctually try to recreate the "scene of the crime" ("Wait. You were standing over here, and I was holding your car keys…") or to recall what was being said before or just after the missing element.

For the vast majority of us, our memory issues aren't about storage capacity—they're about recall. Our brains contain detailed representations of objects we've seen—for example, one study showed that people can look at thousands of photos and identify, even hours later, which ones they've seen from a large group of ones they haven't. But recalling a specific memory from all of these patterns is the tricky part, and it's one reason we are always searching for helpful tricks and memory hacks of one kind or another.

Why Are Some Memories So Painful?

Many memories are marked for storage in part because of the emotion they generated—it's one way the hippocampus sorts the emotionally connected memories from the unemotional—and that emotional trigger is encoded into the memory itself. An event with any sort of emotional component is more likely to be stored, and more easily retrieved. And as you relive

Short-Term Memory Test

How many items can you store in short-term memory? Read the first line in the chart below, then close your eyes and repeat it back. Go on to the second line. Stop when you can't remember. That's the limit of your short-term memory.

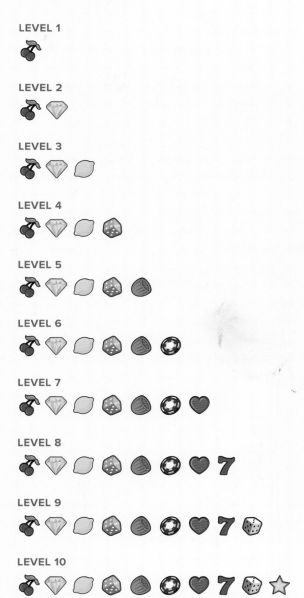

LEVEL 1

LEVEL 2

LEVEL 3

LEVEL 4

LEVEL 5

LEVEL 6

LEVEL 7

LEVEL 8

LEVEL 9

LEVEL 10

When asked to remember a too-long list of items, people tend to remember the first few items and the last few items the best.

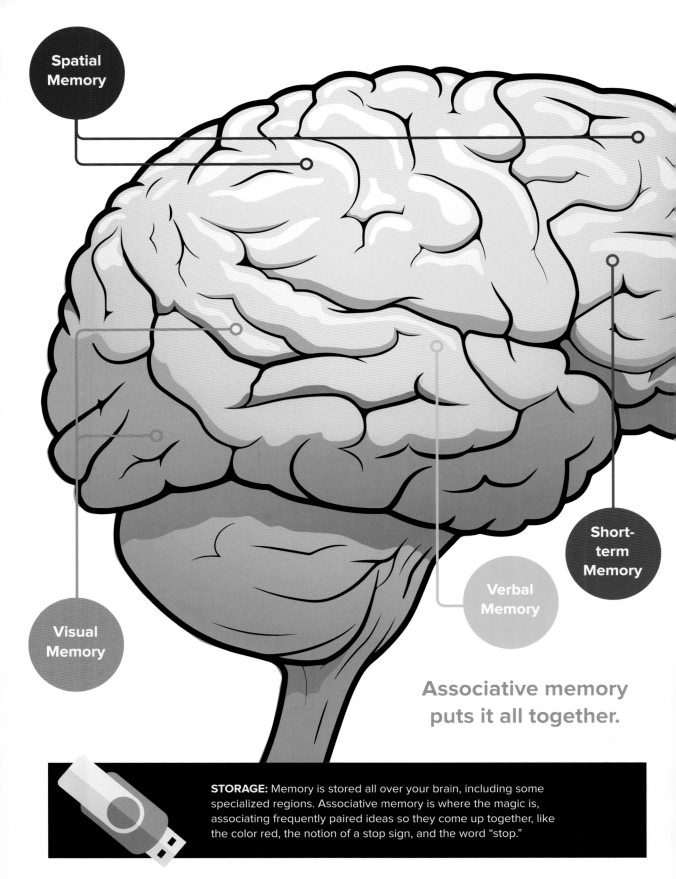

Spatial Memory

Visual Memory

Verbal Memory

Short-term Memory

Associative memory puts it all together.

STORAGE: Memory is stored all over your brain, including some specialized regions. Associative memory is where the magic is, associating frequently paired ideas so they come up together, like the color red, the notion of a stop sign, and the word "stop."

the memory of an emotionally charged event, you are vulnerable to reliving those emotions, even years later. Think about deaths, break-ups, or divorce. When you want to wallow, all you have to do is play that guilty-pleasure song that always makes you tear up.

Sometimes, painful or terrifying memories can be blocked or locked away from the conscious mind, either purposefully (through **suppression**) or subconsciously (through **repression**). When you refuse to deal with something, or say "I don't even want to think about that right now," that's suppression. In one experiment, participants who lied to improve their outcomes during a coin-toss game were able to willfully forget their bad behavior, in what the study's authors called "unethical amnesia."

Repression is different: It involves pushing a very disturbing memory or idea so far out of consciousness that you can no longer recall it at all, and yet can still feel its effects, which may now feel mysterious. Repressing memories is neither a pinpoint affair nor easy on the system—it can impair visual perception and the ability to form visual memories, and it can cause existing, non-targeted memories to get erased along with the target memory. It can harm your ability to form new memories by preventing the stressed hippocampus from fully encoding new memories for storage. One student who survived the Columbine shootings was unable to remember anything she learned in class in the weeks after the event. PTSD sufferers, by and large, find it difficult to form new memories.

The Incredible Falseness of Memory

The attacks on New York, Pennsylvania, and Washington, DC on 9/11 traumatized the world, searing the shocking images of planes hitting buildings into the memory of millions of people. If ever an event could be counted on to be remembered correctly, imbued with emotion and reinforced by shared and repeated experience, surely this would be it.

And yet...

Forty percent of people misremember *what they themselves did that day*. Right after the attacks, researchers asked more than 2,000 people to share their personal experiences. Then they checked in on the respondents: one, three, and 10 years later. 4 out of 10 people falsified their own stories—presumably unconsciously—misremembering details like whether they were in their office or on the street when the plane hit the second tower. (Another study found many claiming they'd viewed the wreckage of the plane in Pennsylvania in news reports on 9/11, even though that footage wasn't shown until the day after.) Interestingly, once they'd falsified their own story, subjects tended to stick with the new version, even many years later.

Here's what seems to be happening: When people retell an event, even to themselves, they unconsciously try to make sense of it by smoothing the edges and filling in details in a logical fashion. "You begin to weave a very coherent story," said the author of the 9/11 study, Dr. William Hirst. "And when you have a structured, coherent story, it's retained for a very long period of time." Hirst even suggested, in a *Time* article, that this effect could explain Brian

Why do we talk to our pets?

Anthropomorphism is actually an important social-cognitive trait. Our medial prefrontal cortex and left temporoparietal junction help us build bonds by attuning to and empathizing with one another's personalities, thoughts, and intentions. This instinct carries over to non-humans, especially those that have faces and/or behave enough like us to warrant further attention and care, which is why it's so tempting to believe that your barking dog actually "prefers" this date outfit to the last you tried on. The notion is so satisfying that we also attribute human qualities to plants, boats, hurricanes, and even the stock market, to help make sense of occurrences that are as inexplicable and unpredictable as we are. Think about what's actually going on in your head, next time your "cranky" car won't start, or when the !$@?&#! slot machine "denies" you that third cherry.

"If you are motivated to try and prevent yourself from reliving a flashback of that initial trauma, anything you experience around the period of time of suppression tends to get sucked up into this black hole as well."

–Dr. Justin Hulbert

NEVER FORGET...
When a story's told many times over, like where you were during the attacks on 9/11, people feel the need for a coherent story, and fill in details on later retellings that may not be correct.

Williams' infamous telling of the false narrative about his work in Iraq: Because the new story seemed to make logical sense, his mind convinced him it was the true story.

Every time we remember something, we are reconstructing it—and there's danger of rebuilding it incorrectly. In fact, retelling a story over and over is a great way to get it wrong— we tend to do a better job getting the details right on anecdotes we *haven't* retold over and over again, presumably because each step between the original event and *this* retelling is like a game of telephone that keeps diverging.

It's disturbingly easy to implant completely false memories, as psychologists have shown again and again. When presented with misinformation or asked leading questions, a nervous subject can form a "memory" they firmly believe is their own. John Bradley, a decorated Navy sailor, toured the world as a hero (raising money for war bonds) after raising the flag over Iwo Jima in that iconic Pulitzer-Prize-winning photo. But the Marines revealed in June of 2016 that Bradley wasn't in the photo at all—

he'd taken part in an earlier and completely separate flag-raising. Did he truly believe he was in the photo? It's possible that stress, repeated confirmation from outside sources, and even his own storytelling all reinforced the false memory and convinced him he was actually there.

People with false memories often believe them just as strongly as if they were real. And why not? Remember that a memory is nothing more than reliving a pattern, and there's no reason to think that reliving a "false" pattern feels different from reliving a true one. Your memory can be just as vivid, even if the event never happened, and your confidence in *all* your memories is at least a little misguided. "It's not whether our memories are false, it's *how* false they are," says Julia Shaw, author of *The Memory Illusion: Remembering, Forgetting, and the Science of False Memory*. "Our personal past is essentially just a piece of fiction."

The Eyewitness Conundrum

So let's put all this together. People's memories

are inherently faulty, they're usually unaware of that fact and trust their memories, and it's easy to implant false memories, consciously or unconsciously. Now imagine you're on trial, accused of a murder you didn't commit, and will have to rely on eyewitness testimony to win your case.

Suddenly the point isn't so academic.

We have acted, for most of the course of human civilization, as if eyewitnesses have recorded the event in their heads and can simply play it back for the courtroom. Today, we know with absolute certainty that memory is nothing like that—and yet, eyewitness testimony still forms the backbone of our criminal justice system.

Eyewitnesses can and do unconsciously fill in details that seem to make sense to them, or change facts to make a story more cohesive or compelling. Their personal testimony can be compromised by "leakage" from other reports they hear; and they can be easily influenced by others, particularly authority figures like police officers and attorneys, adjusting their story to better fit leading questions or to include details supplied consciously or unconsciously by a third party hungry for a conviction.

Unusual Memory Phenomena

The organic, faulty system we evolved to create and retrieve memories can produce some mysterious memory conditions. Here's science's best understanding of what's going on.

Why can't you find a word that's "on the tip of your tongue"?

This issue with recall, also known as blocking, involves what's sometimes called **metamemory**—where you know that you know something, but can't remember it exactly. It happens because your brain organizes information in a dispersed manner, which can make recall difficult. Triggers can definitely help pry out the hidden word; in this case, try running through the alphabet to find the starting letter of the word you're looking for.

I can drive "on autopilot" when I know the way. How's that possible?

This is an example of your brain's procedural memory—the same part that lets you easily reprise repetitive tasks like typing and tying your shoes. When you've logged thousands of miles of driving time, the dozens of small things you do while driving (engaging turn signals, changing foot pressure on the gas, etc.) become rote and automatic, freeing your conscious mind up to think of other things. It's potentially dangerous, of course, so two hands on the wheel.

What's "déjà vu"? And why does it happen?

Déjà vu (French for "already seen") is when something seems eerily familiar—a view, a conversation, a feeling, a paragraph in a book—though, in fact, it's a brand new experience. Technically, déjà vu only refers to things seen, but people commonly use the phrase across all kinds of experience. Déjà vu might

71% of all wrongful convictions later exonerated by DNA evidence involved eyewitness misidentification.

Justice Denied: The Tale of Marvin Anderson

Marvin Anderson was 18 years old when he was convicted of abduction, rape, sodomy, and robbery. Twenty years later, Anderson was exonerated—proven not guilty by DNA evidence and technology not widely available at the time of his original trial. He'd been convicted based almost entirely on eyewitness testimony, which we now know to be incredibly flawed. The miscarriage of justice against the innocent Anderson included:

- Anderson was brought in as a suspect only because the perp had reportedly told his victim he "had a white girl," and the cop thought Anderson, who lived with a white woman, fit that vague description.

- The victim—the sole eyewitness—was shown photos of seven suspects, in which Anderson's photo was in color, and stood out from all the others, which were in black and white.

- The victim was next shown a lineup, including Anderson (whose color photo she'd just seen), but nobody else from the photos.

- Anderson's lawyer ignored pleas to call to court a more likely suspect, John Otis Lincoln, who'd stolen a bike later used in the assault.

- Even after this new suspect actually CONFESSED to the crime six years later and offered details in court proving his involvement, the judge—the same as from the original case—refused to overturn Anderson's false conviction.

- When Anderson's lawyers pleaded to have crime scene evidence tested for DNA to prove his innocence, they were told the rape kit had been destroyed.

- When some evidence was found after all—thanks to intervention from The Innocence Project, an organization dedicated to overturning wrongful convictions—requests for DNA testing were still, repeatedly, denied.

The Innocence Project eventually won the right to DNA testing and was able to establish Anderson's innocence and win him a full pardon from the governor, but only after he had already served the full 15 years in prison and another four on parole.

This horrible story is unfortunately not an outlier. The Innocence Project has found eyewitness misidentification to be the single most important contributing factor in wrongful convictions that are later overturned by DNA testing.

happen when a sensory input to consciousness is mistakenly treated like a memory input. Alternately, it could be that you really do know something but can't quite place what it is: You might, for example, experience déjà vu while seeing a portrait of George Washington in a museum and never realize that the image you're seeing is the same portrait you've seen smaller, in green-and-white, a million times on the one dollar bill. To test this idea, researchers showed subjects dozens of 3-D scenes (of hotel lobbies and the like), restaging some scenes exactly but with different variations—a rough similarity that went unnoticed by the conscious mind, but induced feelings of déjà vu in some subjects.

Why did you forget what you just came into this room for?

This is called the **Doorway Effect**. Studies have shown that a person given a task that takes him from one room into another forgets the task more often than one who completes the task across the same distance within a single room. It's surmised that the act of crossing a threshold into a new room—which invites the eye to assess a new geographical situation for threats and opportunities—automatically repurposes some of your short-term memory, erasing what was there and making you more likely to forget what you came for.

What's "déjà vu"? And why does it happen?

Déjà vu (French for "already seen") is when something seems eerily familiar—a view, a conversation, a feeling, a paragraph in a book—though, in fact, it's a brand new experience. Technically, déjà vu only refers to things seen, but people commonly use the phrase across all kinds of experience. Déjà vu might happen when...okay, yes, we repeated this one on purpose. Just making sure you're paying attention.

Why can't you remember the name of the person you just met?

This one's called the **Baker Effect**. If someone tells you she's a baker, you can connect it to things you already know—the smell of bread, the puffy white hat. But if she says her *name* is Baker, your brain has nothing tangible to associate it with, and an isolated idea is harder to recall than one with built-in connections in your brain. Picture her making you donuts, and you may find her name easier to recall.

Is amnesia a real thing?

Amnesia, the total loss of some or all memories, *does* happen, but it's very rare—and the really fun soap opera version, where the person forgets not just their memories but also their identity, is rarer still. Amnesia can be caused by physical factors like injury or dementia, or by psychological factors like repression. Most cases of amnesia are temporary and resolve themselves without intervention. Interestingly, people with amnesia often find it hard to imagine the future; thinking of the past and of the future light up the same brain areas: the **prefrontal cortex**, medial temporal lobe, and the retrosplenial cortex.

The Silver Lining of Forgetting

For all this talk about remembering and recalling, it's easy to overlook how important it is to *forget*. Scientists are beginning to rewrite the common narrative of forgetting from curse to blessing. Forgetting keeps our memory and problem-solving skills in tip-top shape by filtering out extraneous information, keeping our noodles nimble enough to absorb new input and focus on what's currently relevant. Early research has identified a protein called musashi that may help manage this process.

Does "Photographic Memory" Exist?

Can some people remember everything that happened to them in vivid photographic detail? The short answer is no—your crime fiction has been lying to you again. But there certainly are some people with better than average memories. Those with **hyperthymesia**, for example, have exceptional **autobiographical memory**, whether they want it or not. Their past plays alongside the present, like a split screen of two movies playing at the same time—and has been described as a living nightmare. It seems that the amygdala, which codes some memories with special "save me" significance based on emotional or social relevance, are overactive with hyperthymesiacs, and they get *too* many memories. A similar condition is **eidetic memory**, where a vivid **afterimage** of a memorable situation lingers in the mind's eye for up to a few minutes before fading away. The character Sheldon in *The Big Bang Theory* experiences this.

Closing your eyes can help you remember better. In one study, people who'd just watched a film were asked to remember details. Those who were told to close their eyes before remembering answered 23% more questions correctly.

THINK ABOUT THIS

At the other extreme, scientists have identified a new and mystifying condition: severely deficient autobiographical memory (SDAM), the inability to form episodic memories. These memories are the who's, what's, when's, and where's of your life experiences starring you as the protagonist—and SDAM sufferers can't form them at all. Susie McKinnon, the first of only a handful identified with the condition, has no memories—no mental stories—of her wedding or any other significant event in her life. She can flip through her wedding album and recite stories she's learned from her husband, but the photos themselves feel, to her, like someone else's. Surprisingly, she and others affected by SDAM live otherwise normal and fulfilling lives, enjoying happy relationships, working steady jobs, and having excellent—even above average—semantic memory for facts and concepts. This helps McKinnon remember quite well the music she sings in choir, despite having no memory of ever performing.

What's the Deal With Savants?

Savants, those who possess particular and remarkable talents despite mental and/or physical disability, by and large, can shuttle information readily from short-term memory to long-term memory (i.e., "Just file everything!"), resulting in the ability to pull off incredible mental feats like being able to memorizing the entire phone book, or recalling a city well enough after a single helicopter ride to draw a detailed panorama. Kim Peek, the real-life inspiration for Dustin Hoffman's "Raymond Babbit" character in *Rain Man*, was not autistic but was born without a corpus callosum, the connective tissue between the brain's hemispheres. He could remember an incredible level of detail about recent events (e.g., Babbit's recollection of 246 toothpicks), but couldn't organize experiences or make judgment calls; his reasoning and verbal abilities never progressed beyond those of a 5-year-old.

Memory Decline

Like many cognitive functions, memory also seems to decline naturally with age, possibly in part because regions of the brain involved in memory shrink, and blood flow to the brain slows with time. Older memories tend to survive better than newer memories, and the ability to form new memories can worsen. So if Grandpa forgets to come to your rehearsal dinner because he was busy telling the usher his painfully detailed Korean War story, don't be insulted—this is normal. Memory in elderly people can be compromised by other factors, too, such as when a decline in hearing makes it harder to hear a new person's name, or when loneliness or **depression** dim your motivation to learn new things. Memory loss can also be exacerbated by alcohol use (by slowing blood flow to the brain) and marijuana (where long-term use has been shown to affect verbal memory).

Brain Boosting: How To Keep Your Memory Strong

What use is living an awesome life if you can't remember all the details? Here are some things you can do to keep your memory function strong as the years go by.

Eat a brain-smart diet. The Mediterranean diet, in particular, and specific foods like salmon, blueberries, and cocoa can help brain health generally. Avoid diets high in cholesterol and fat: You already know they're bad for your heart, but new evidence suggests (though it's controversial) that they can hurt your noodle too, creating sticky protein clusters called beta-amyloids that clog the brains of Alzheimer's sufferers.

Exercise. Exercise generally conveys a lot of positive brain-boosting benefits, including increasing blood-flow to the brain, reducing inflammation, and helping stave off depression. Regarding memory in particular, regular aerobic exercise has been shown to boost the size of your memory-critical hippocampus—other kinds of exercise didn't show this effect. And exercising four hours after learning something boosted retention of the learned material by 10% (but there was no effect when exercising right after studying). After-school soccer, anyone?

Get enough sleep. A recent study indicates that your mom was right: You DO need more sleep. According to UC Riverside psychology professor Sara Mednick, in an article for *CNN Health*: "Sleep helps transform short-term memories into long-term memories by helping make stronger connections between these new experiences and our old memories, that allows the new experiences to be integrated with our general knowledge and understanding of the world."

Don't expect too much from nootropics (so-called "smart drugs"). There isn't a lot of evidence yet that cognition-enhancing drugs, such as those designed to boost your brain's memory-transmitting **acetylcholine**, do much to improve memory. It may be simply that they put the brain in an optimal chemical state to do what it already

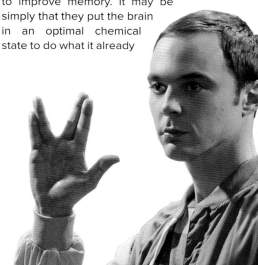

> *Photographic is a misnomer—I have an eidetic memory, as I've told you many times before. Most recently last year, during lunch on the afternoon of May 7th. You had turkey and complained it was dry.*
>
> –Sheldon Cooper, *The Big Bang Theory*

FIVE AWESOME
MEMORY HACKS

A memory system that evolved in the natural animal world isn't always the perfect tool for remembering the minutiae of our modern digital civilization. But you can leverage your natural assets by converting what you need to remember into something your brain knows how to remember.

To temporarily remember a list of random items, like a grocery list

Hijack your spatial memory with the **Palace Technique**: Your memory's well equipped to record spatial data, like hunting grounds and watering holes, so turn your grocery list into a story that plays out across a physical space. For eggs, milk, bread, and apples, think "I walk out to my driveway, where my car's been egged! I see a pool of milk at the curb, and a loaf of bread stuck in the mailbox, as I turn right, and there's a new apple tree in the Johnsons' yard..." Sounds stupid? Try it and be amazed.

To remember a small complete set of things forever

Give yourself a permanent recall trigger with acronym mnemonics. Remember HOMES to help recall the Great Lakes: Huron, Ontario, Michigan, Erie, and Superior? Words can be hard to recall in a vacuum, but a simple trigger can get you closer—in this case, knowing the first letter of the five words you need. You're remembering one word instead of five, and because you feel clever you might even get an emotional boost that aids long-term memory.

To remember the name of a new person you meet

Hack your visual memory—which is stronger than your memory for words—by picturing the new person's name written on their forehead with magic marker in your favorite color. Alternately, tie him to a famous person you already know by picturing them as that person. (So for a new Michael, maybe you picture them bald and dunking like Michael Jordan.) However you get there, storing his name as a visual memory will make it easier to recall than storing it as a verbal memory.

To temporarily remember a bunch of facts, like for a test

Divide whatever time you have to cram into two separate study sessions, with a break in between. Studies have shown you learn twice as much in the same amount of time if you divide that time in half and the second session revisits the first...repetition teaches your mind that something's important, and hence worth remembering. Other unconventional study tips that could help: Sing the words of paragraphs you're reviewing, and touch individual fingers when memorizing short list items.

To remember a string of numbers

Convert them to words and tell yourself a story. With no emotional or physical component, numbers are hard to remember. The **Major System** pairs each digit with particular consonant sounds, so 1 goes with "d," 7 with the "k"/hard "c" sound, and 6 with "shh"; you then string these together with any vowels you choose. So 1776 is "duck cash," and you could picture a duck stealing money from the desk where the Declaration is being signed. The sillier the idea, the easier it is to remember.

IS "STATE-DEPENDENT LEARNING" A THING?

Yes, for the most part. After rats with a certain drug in their system learned to escape a maze, for example, they could only escape it when they were on the drug.

Another study of college students demonstrated that test takers should repeat study conditions at test time where possible—it's far better to be sober for both, but if you couldn't avoid being drunk when studying, you're marginally better off being drunk for the test, too.

does well. In fact, the caffeine and other stimulants they often contain can produce similar effects on their own: They don't improve your brain's ability to memorize, they just make it work more efficiently in the here and now.

Helen Keller was born with sight and hearing, but lost them at around a year and a half due to illness. It was a dark and silent world, and she was mostly uncommunicative, even when her patient teacher Ann Sullivan signed words like "doll" and "mug" into her hand. And then, one day, something clicked: When Sullivan signed the word for "water"—pressed into one hand while water was being poured over the other—it reminded Helen that she had once known a word for water. Suddenly she remembered that *words mean things*, and (at least in the theatrical and movie version, *The Miracle Worker*) flew about her house and yard with her teacher in tow, trying to absorb the words for all the things in her life.

It was memory that led Keller back to a world where things had meaning. That's what memory does for us, too: It adds meaning to experience, registering emotions and physiological responses right there along with the details of whatever's being encoded. It helps us lean into positive things we've already experienced and shy away from negative things. The process is inexact and squishy: The same neurons are used for multiple patterns, and the patterns themselves leave an electrochemical signature that can grow faint with time.

But the processing of meaning doesn't end with encoding items into memory. Our brains continue processing the new information—forging new connections, whittling away what is unimportant—even when our conscious mind isn't paying attention. This is largely accomplished through a crazy system wherein consciousness shuts down but the brain stays active, shifting into a different mode where outside sensory influences are minimized and memories and thoughts play out for hours. In this state, the brain entertains itself with realistic visions of probable and improbable things, for various subconscious purposes.

You know the process as sleep and dreaming—the subjects of the next chapter.

Are our memories getting worse because of smartphones?

Now that we regularly use smartphones to retrieve things we used to have to remember, like who won the 1996 Super Bowl and how to get to Uncle Bob's house, are we impairing our ability to memorize things? It's a little too early to say definitively, but the early returns seem to indicate yes. In one study, students in a museum were asked to photograph some objects and merely observe others; they remembered fewer details about the ones they photographed. In another, participants were asked to navigate to a known location. Those who relied on GPS to get them there had less activity in their memory-forming hippocampus than those who toughed it out without.

BRAIN FOOD **MADE EASY**

Feed your brain what it *needs*— not what it *wants*.

If you're looking to nurture your noggin, everyday foods can help to improve memory, focus, and can even spur the growth of new brain cells.

FRIES

AVOCADO

DARK CHOCOLATE

BLUE-BERRIES

SUGARY JUICE

CHEESE

WHITE RICE

WALNUTS

SALMON

TUNA

CHAPTER 4

SLEEP, DREAMS, AND THE SUBCONSCIOUS WORLD

"To sleep: perchance to dream..."

Shakespeare's Hamlet is not afraid of dying—he's afraid of *dreaming*. In his signature "To be or not to be..." soliloquy, he reveals that the only thing staying his hand from suicide is the possibility of a hellish eternity of bad dreams.

A sobering thought indeed.

Sleeping and dreaming have fired our imagination throughout the course of human civilization. Snow White and Sleeping Beauty are put in enchanted sleeps; Juliet enters into a chemically induced coma. (Sadly, Romeo didn't get the memo.) Adam was asleep when God extracted a rib to make Eve, and Samson slumbered while Delilah gave him that fateful trim. Rip Van Winkle slept for 30 years, missing the Revolutionary War, and King Arthur snoreth even now, in Avalon, waiting to be summoned again.

What a strange state sleep is when you think about it. Otherwise perfectly functional humans, by the houseful, suddenly become root vegetables: they close their eyes and go quiet all night, alive but unaware. (It has to be pretty damn helpful, from an evolutionary standpoint, if it's worth introducing the insane danger of being quiescent and vulnerable for hours.) We know it's critically important—and yet, despite decades of passionate study, we still don't completely understand the function of sleep, or dreaming, or consciousness itself.

But we've learned a lot. And it's fascinating.

MIND BENDERS

40 *Percent of adults who are sleep-deprived*

1 *Percent of adults who sleepwalk on a regular basis*

6 *Years of your life you spend dreaming*

Pregnant women who have nightmares during pregnancy have easier births than those who don't.

sleep provides, our bodies and brains simply can't continue to function. Sleep-deprived rats lose immune function and die in a few weeks; sleep-deprived humans start to hallucinate, and some have seizures, after just a few days.

What Makes Us Sleepy?

There are two systems that work to regulate your sleepiness: **circadian rhythms**, which coordinate your sleepiness and wakefulness with natural sunlight and darkness, and **sleep homeostasis**, a metabolic process that occurs throughout the cumulative hours you've spent awake.

Your circadian rhythms, centered in your hypothalamus, evolved to cleverly align the times when you feel sleepy, awake, and hungry to external stimuli, like sunlight and darkness. Your peak "naturally sleepy" times are roughly from 2-4 AM and from 1-3 PM, so that's when you're likely to get drowsiest. But sunlight and darkness are important inputs, and there's evidence this system has been severely compromised by our modern propensity to stay up way past dark and rely on artificial light.

These unnatural sleep habits can be corrected by the second system, sleep homeostasis. This process involves a neurotransmitter called **adenosine**, which is produced and captured by specialized receptors the whole time you're awake; gradually over the course of the day, as adenosine accumulates in the receptors and there are fewer places for new adenosine to bind, your body becomes sluggish and feels "sleepy." When you crash at last, the adenosine is slowly cleared away, such that if you get enough sleep, you should awake feeling refreshed. Think of it as a sleepiness impulse counterbalanced by an arousal impulse, chemically connected to seesaw back and forth over the course of the day/night. When you lie down to sleep, the sleepiness impulse reduces activity in the hypothalamus, thalamus, **basal forebrain**, and **frontal cortex**; in the morning the arousal impulse gently takes control and reverses the chemical flow so you wake up.

Why Do We Need Sleep?

Sleep seems to help with a number of important activities, including re-energizing the body's cells, rejuvenating the brain, consolidating memories, and regulating various cognitive functions. Sleep correlates with real changes in the structure and organization of the brain. And we can get a sense of just how important sleep is by paying attention to what happens when we don't get enough of it. Falling asleep while driving causes an estimated 100,000 automobile accidents a year and around 1,500 deaths. The driver of a Manhattan-bound commuter train that derailed in the Bronx in 2013, killing four, was asleep at the wheel—so was the driver of the 2003 Staten Island Ferry crash that killed 11. Some 200,000 workplace accidents a year are blamed in part on sleep deprivation; it was cited as a contributing factor in the Challenger explosion and the Three Mile Island and Chernobyl nuclear disasters and the wreck of the Exxon Valdez that dumped 10 million gallons of crude oil into Alaska's Prince William Sound. Without the healthy reset

Women dream about both genders in roughly equal proportions, but **70% of the characters in men's dreams are other men.**

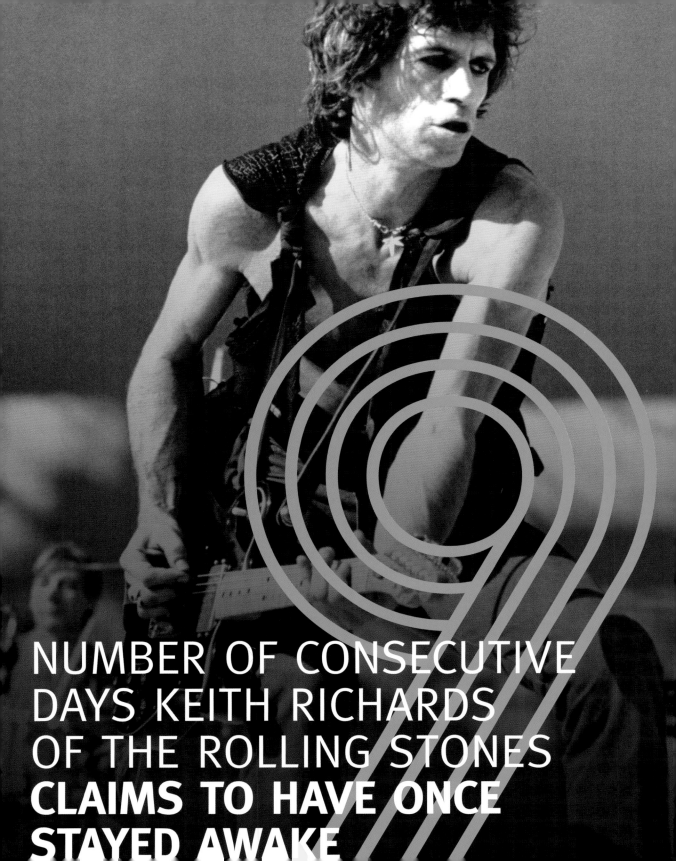

NUMBER OF CONSECUTIVE
DAYS KEITH RICHARDS
OF THE ROLLING STONES
**CLAIMS TO HAVE ONCE
STAYED AWAKE**

What Happens When You Sleep?

If you sleep long enough in a given night, every 90 minutes to two hours you will typically cycle through four distinct sleep stages. Here's what happens, as explained by WebMD:

FIRST STAGE: NON-REM (RAPID EYE MOVEMENT), OR NREM1 SLEEP

We all know this as the "nodding off" stage. In this stage, which typically lasts about 5 to 15 minutes, your brain's electrical activity begins to quiet down. Your eyes are closed, but you can be easily awakened.

SECOND STAGE: NREM2 SLEEP

This stage is characterized by light sleep, and lasts around 30-60 minutes. Your body temperature drops and your heart rate slows. Your brain's electrical activity simmers down further into the onset of deep sleep. The sandman is digging in.

THIRD STAGE: NREM3 SLEEP

This is the deep sleep stage, lasting around 20-40 minutes. Here, your brain is the closest it ever comes to total rest: Your metabolic rate slows down and some neurons stop firing. You are harder to rouse, and can become disoriented if awakened. If you're going to sleepwalk, this is when it happens. A lot of the night-shift work of the brain takes place now: strengthening the immune system, regrowing tissue, repairing DNA, and building bone and muscle. It's also in this stage that we start to forget the unimportant, thanks to **synaptic pruning**—synapses are gradually weakened in connections

deemed unimportant, freeing up mental capacity for the next day's new memories.

FOURTH STAGE: REM SLEEP

This is the time when most dreams occur. Your brain is as active here as when you're awake, as the basal forebrain pumps out the neurotransmitter acetylcholine, your brain's electrical activity quickens, and your eyes zip around like pinballs beneath their closed lids. Yet all this neural activity is mediated by the neurotransmitter **Gamma-Amino Butyric Acid (GABA)**, which keeps your mind unconscious and your body still. It's been hypothesized that acetylcholine facilitates learning and memory processing here, connecting what you've

SLEEP PATTERNS OF THE RICH AND FAMOUS

Sleep styles are a deeply personal thing, and some of Earth's highest achievers reportedly had—and have—some unusual bedtime habits.

Charles Dickens: Always slept facing north

Nikola Tesla: Never slept more than two hours a night

Winston Churchill: Took a two-hour nap every day

Emily Brontë: Walked in circles until she could fall asleep (she was an insomniac)

Mariah Carey: Sleeps 15 hours a night, in a bed surrounded by humidifiers

Michael Phelps: Sleeps in a low-pressure chamber that maximizes his breathing capacity

Frank Lloyd Wright: Woke up at around 4 AM, worked for three or four hours, then went back to sleep

ZZZZ People taking anti-depressants like Prozac, Zoloft, and Paxil are more than twice as likely to sleepwalk.

learned and experienced throughout the day to what you already know. It's also the stage in which procedural memory (for example, playing poker or piano) seems to be consolidated. The science, however, is still unclear. (See Lucid Dreaming on page 78.)

It takes 90 minutes to two hours for the cycle of stages to complete, so it is possible to experience three to five cycles in a given night. If your cycle gets interrupted and you have trouble getting back to sleep, read on for some tips later in the chapter.

Nothin' But Nap: The Strangely Satisfying Uberman Sleep Cycle

Don't like the idea of committing to sleeping seven whole hours in a row? Consider trying the **Uberman Sleep Cycle**, which replaces the sleep-all-night idea with a 24-hour succession of well-spaced power naps that total just two to three hours of sleep per day. This is a specific example of **polyphasic sleep**, or sleeping in multiple phases, and there's some evidence that humans' natural state, before the advent of artificial light messed it all up, was polyphasic: two blocks of sleep with an hour or so of wakefulness in between. There's controversy over whether adopting such a pattern is sustainable over the long haul, but with overachievers like Leonardo da Vinci, Thomas Edison, and Nikola Tesla reportedly having been polyphasic sleepers, it's tough to dismiss. If you want to try, prepare for a brief sleep-deprivation learning curve, and prepare to spend

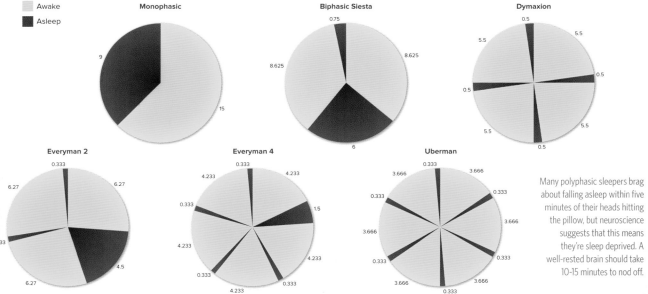

Many polyphasic sleepers brag about falling asleep within five minutes of their heads hitting the pillow, but neuroscience suggests that this means they're sleep deprived. A well-rested brain should take 10-15 minutes to nod off.

When Sleepwalking Has a Body Count

In 1987, Canadian Ken Parks got in his car in the middle of the night and drove 14 miles to the home of his in-laws. He then took a tire iron out of the car, let himself into their house with his key, and went into the couple's bedroom, where he strangled his father-in-law into unconsciousness, then beat his mother-in-law with the tire iron and stabbed her dozens of times with a kitchen knife. After she was dead, Parks got back in his car and drove himself straight to the police station. He later stood trial, and after some impassioned arguments, was found not guilty by reason of "non-insane automatism"... in this case, known as sleepwalking.

Parks' life at the time of the murders was a recipe for stress: He was unemployed, with a wife and infant child to support, was suffering from insomnia and depression, and had a severe gambling problem he'd fed with $32,000 stolen from his employer, who fired him on discovering the crime. When he showed up at the police station after the killing, Parks seemed in sincere distress, repeatedly confessing variations of "I've just killed two people with my hands," having thought incorrectly that he'd killed his father-in-law too. He was apparently not in pain, although in the process of the beating and stabbing he'd cut tendons in both hands. Parks' EEG scans showed he had unusual activity in deep sleep consistent with parasomnia and the jury ultimately acquitted him of all charges, declaring him neither guilty NOR innocent by reason of insanity—rather, they pronounced the killing to have occurred thanks to a very particular set of external circumstances unlikely to ever happen again. No jail, no mental hospital confinement. He was simply free to go.

Parks was an early, but high-profile case of homicide-by-sleepwalking; there have been at least 68 such cases. Interestingly, he was due to have had, on the day of the murder, an uncomfortable conversation with his in-laws, at a family barbecue, about his unemployment and his upcoming trial for fraud. And his managing of the complexity of the midnight drive to their house (including multiple unfamiliar intersections) and of the crime itself (multiple murder weapons and victims) is beyond what most experts believe sleepwalkers are capable of. Finally, it's generally no harder to wake a sleepwalker than anyone else, casting a lot of doubt on the judgment that he somehow slept his way through not one but two prolonged violent acts.

a lot more time at 7-Eleven and 24 Hour Fitness.

Are There Really Natural "Early Birds" and "Night Owls"?

Scientists say that there's something to these **chronotypes** of people naturally better suited to early mornings or late nights, though it's not as cut and dried as our colorful idioms imply. "It is a continuous trait, as is body height or shoe size," says circadian rhythms expert Till Roenneburg of Ludwig Maximilian University, on LiveScience.com. "There are very, very short people, very, very tall people, and the rest are in between." One interesting study of self-described "early birds" and "night owls" found that while there were no statistically significant differences in health or intelligence, night people tended to be slightly wealthier and have more sex partners, while morning people tended to be slightly happier (though it's hard to imagine why, with all those rich night people cavorting around until three in the morning).

What is Sleepwalking?

When you're asleep, your brain is very active—at times as active as it is when awake. But your brain paralyzes your muscles, to make sure you don't carry out whatever craziness you're thinking of. This is done by releasing a neurotransmitter called GABA that inhibits the motor system; essentially, it makes sure the body ignores messages from the mind, freeing the brain to do its thing. **Sleepwalking** (which falls under **parasomnia**, the umbrella term for normally wakeful activities engaged in while technically sleeping, like sleep talking, nightmares, and sleep aggression) happens when the gate is leaky, and the body listens to the mind even though the mind isn't really paying executive-function level attention, with often harmless but sometimes disastrous consequences.

Sleepwalkers' eyes are often open, because the part of their brain responsible for motor function is awake, and they can receive sensory input. The cortex, the part responsible for higher-order thinking and decision-making, is asleep, but the sleepwalker can act on signals received from the dreaming brain. So they can't do very complex things or tackle new challenges, but they can often accomplish the kinds of tasks they've done a million times before. One Scottish chef has been caught whipping up late night spaghetti Bolognese while sleep-

HONEY? In a survey of 5,000 people, the dream that haunts the most people is a dream about a partner's infidelity.

walking; sleepwalkers have walked off subway platforms and into rivers, had sex with strangers, texted ex-girlfriends in mortifying detail, painted elaborate art, and shot themselves. Sleepwalking has even been blamed for homicides—dozens of them—though whether that's legally excusable is controversial. (See *When Sleepwalking Has a Body Count* on the opposite page.)

Sleepwalkers fall down and get hurt when they sleepwalk on vacation, because they can't maneuver in the unfamiliar environment. They typically don't feel pain from the injury, though, until they wake up. The idea you "shouldn't wake a sleepwalker," by the way, is poppycock—you could very likely save a life, especially if the person is stumbling out to the road. Just be gentle, as they'll likely be startled and disoriented. Resist the temptation to publish their plight as a hilarious YouTube video.

Other Sleep Disorders and Conditions

INSOMNIA

Insomnia is the inability to fall asleep, to return to sleep after a disruption, or to stay asleep even when you're exhausted. Doctors recognize two types of insomnia: **acute**, as when you're stressed about a particular performance review; and **chronic**, when you suffer insomnia at least three times a week for at least three months, usually indicative of an ongoing medical problem like anxiety and/or depression. Chronic insomnia reportedly affects 10% of Americans.

Insomnia is associated with increased metabolic activity in the prefrontal cortex (the area just behind your forehead). Usually it slows down when you're tired, and when it doesn't, you feel like you can't relax. An innovative technique—a "cooling cap" that lowers the temperature of the prefrontal cortex—helps about 75% of sufferers fall and stay asleep.

NARCOLEPSY

The simplistic conventional-wisdom view of **narcolepsy**—that sufferers simply fall asleep wherever they are—isn't totally accurate. But it's close. For most of us, **REM sleep**, that sleep state when dreams happen and the body is paralyzed, occurs at only one point in the sleep cycle. But for narcoleptics, the normal pattern is short-circuited, and they can slip into that deep REM sleep at other times throughout the sleep cycle, or even during waking hours. As a

Blame evolution for your struggles falling asleep in a hotel the first night. Your brain's left hemisphere is "staying awake" while anticipating possible threats. Of course, making camp in a saber-toothed tiger den would've been at least as scary as your hotel's breakfast buffet.

People who regularly do "shift work" and/or work across time zones (like EMTs and flight attendants) have higher instances of diabetes, heart disease, reduced fertility, and cancer. Shift work ages your brain and is classified as a probable carcinogen by the World Health Organization.

result, narcoleptics simply can't control when and where they fall asleep.

SLEEPING BEAUTY SYNDROME

At the other extreme end of sleep disorder is Kleine-Levin Syndrome, or **Sleeping Beauty Syndrome**, whose sufferers can't stop sleeping, sometimes for weeks or months, except to eat or go to the bathroom. As appetizing as that might sound to those of us stressed from overwork, it's actually quite a curse, according to folks like Beth Goodier, an Englishwoman who contracted the condition as a teenager. She is disoriented and often speaks in a baby voice in her moments of wakefulness, and ultimately was forced to drop out of school. The condition mostly affects teenagers and the cycle typically lasts more than 10 years before spontaneously burning itself out. The causes of Sleeping Beauty Syndrome are unclear and there's no known cure.

JET LAG

Our carefully calibrated natural sleep cycle is no match for the modern world. The simple act of flying from one time zone to another, for example, wreaks havoc with the cycle. **Jet lag**'s notorious dragging effects aren't a factor of getting too little sleep; rather, they're attributable to the way travel has disturbed your natural clock. Your circadian rhythms, the body's 24-hour biological clock, are designed to make you tired when it's dark and awake when it's light. But if you travel and your 5 AM (when it should be dark) corresponds to local noon, the unexpected sunlight can throw everything off, in what's called a **phase delay**.

It can take up to one day per time zone crossed for your body to get used to the local rhythm if you're going west to east, or about half that time going east to west. The dry pressurization of a plane cabin and staying seated through-

Researchers at Loughborough University found that **women need about 20 minutes more sleep than men per night**, because they tend to juggle more, and hence work their brains harder. As any girlfriend or mother will confirm.

"A dreamer is one who can only find his way by moonlight, and his punishment is that he sees the dawn before the rest of the world."

—Oscar Wilde

You should never wake a sleepwalker. False! Sleepwalkers are easy to wake up, it won't damage them to be gently awakened, and you could stop them from tumbling down seven flights of stairs.

out a flight exacerbate the problem, though staying well hydrated and moving around can reduce the effects. But the bottom line: Jet lag is really, really not good for you. Chronic jet lag can affect memory and cause learning problems. Unlucky hamsters given the equivalent of jet lag showed difficulty, relative to control hamsters, learning new tasks for a month after the jet lag itself, and showed significantly fewer new neurons in the hippocampus, where new memories are born. Flight attendants and others who regularly shift time zones experience a greater incidence of learning and memory problems, and some serious health problems, too, like cardiac disease, chronic bronchitis, and skin cancer.

SOCIAL JET LAG

You don't have to travel to suffer from "jet" lag. There is "jet" lag without the jet: It's called **social jet lag**, and it refers to the discrepancy between your natural body clock—when you're programmed to fall asleep and wake up—and the schedule you actually keep. If the difference between when you have to wake up/fall asleep and when you'd naturally wake up/fall asleep on your own (sans alarm) is notable, you have social jet lag.

Teenagers worldwide are particularly vulnerable to chronic social jet lag, and research shows that their brain development, physical health, and academic performance suffer as a result. Both puberty and earlier school start times are shifting teens' body clocks. A teen with a typical 7 AM school start time needs to fall asleep by 9 PM in order to feel fully rested by her 6 AM alarm, which is tough when she doesn't get out of choir practice until 9:30 PM. Dr. Craig Canapari, director of the Yale Pediatric Sleep Center, says, "This implies three hours of social jet lag on every school day, equivalent to the jet lag of flying from San Francisco to New York five days a week. However, unlike travelling, there is no real habituation."

No wonder over 90% of U.S. teenagers are sleep-deprived.

In the adult world, powering through your sluggishness during the week, then trying to make up the difference on the weekend might make sense on paper, or to a hard-driving boss, but it is definitely out of step with your circadian rhythms. Our sleep hours are no longer driven by whether or not it's dark outside—thanks, artificial light—yet Americans sleep on average just 6.8 hours a day now, an hour less than we did in 1942, according to a Gallup poll. And this social jet lag can have serious health effects. One study found that for every hour of this social jet lag shift, your chances of being overweight increase 33%.

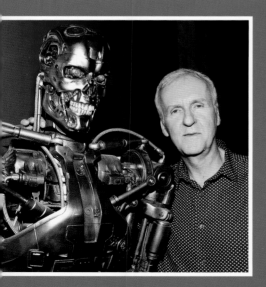

Ten Artistic Works That Reportedly Came to Their Creators In a Dream

- "(I Can't Get No) Satisfaction," by Keith Richards
- *Dracula,* by Bram Stoker
- "Purple Haze," by Jimi Hendrix
- The Terminator robot from James Cameron's *The Terminator*
- *The Persistence of Memory,* by Salvador Dali (a.k.a. "That Melting Clocks Painting")
- *Stuart Little,* by E.B. White
- *Frankenstein,* by Mary Shelley
- "Yesterday," by Paul McCartney
- *Twilight,* by Stephenie Meyer
- *Kubla Khan,* by Samuel Taylor Coleridge

Arianna Huffington's Thrive Global offers tips like reading, meditating, keeping the bedroom dark and tidy and TV-free, and paying attention to when you feel tired.

What's worse than not getting to sleep? Waking up in the middle of the night. If you can't get back to sleep, don't just lie there: Get out of bed and read or do something else low-stress and in low light. *Do not* return those work emails—it's sure to get your brain going, and the unnatural shorter wave light (a.k.a. "blue" light) from your smartphone can suppress the production of **melatonin**, a sleep-and-wake hormone, and shift your body's clock.

Bottom line: Get some sleep!

What Are Dreams?

The benefits of sleep are understandable: It helps rejuvenate our bodies and minds. But what is dreaming and why do we do it? Why do we create and play out experiences in our head, most of which we never remember? What possible usefulness could there be in that? We all seem to have a similar dream experience: A wandering, borderline illogical fictional saga, composed of disconnected memories (often from the day but sometimes from long ago) and new ideas in strange combinations. The fascination with this mysterious process has been with us since the dawn of civilization, reflected in countless human stories. Alice's Wonderland turns out to be a dream, and so do Dorothy's Oz and the world of *The Nutcracker*. In the Hindu religion, Vishnu is a god whose dream is the universe. Or as Morpheus puts it in *The Matrix*: "You've been living in a dream world, Neo." Morpheus, of course, is the Greek god of dreams.

Dreaming largely happens in the REM stage of the sleep cycle, as detailed earlier, which means adults dream about five times per night on average. While dreaming, your brain "fills in the gaps" to connect these juxtaposed ideas into one meandering experiential thread. Dream logic would never fly in the real world ("Wait—how could I have made it all the way to the office wearing a chicken costume and not notice it?") but it makes just enough sense to the active part of your sleeping mind that the flawed logic usually doesn't rouse you to full consciousness. Later, you'll feel strangely compelled to relate your dreams to others, who will politely pretend to be paying attention to your rambling tale.

COMMON DREAMS AND WHAT THEY (MAYBE) MEAN

Dreams don't mean anything, and this isn't remotely scientific. But people record similar dreams across ages and cultures—dreams of flying, for example—and it's irresistibly tempting to try to decipher "what they really mean." *Mental Floss* compiled learnings from four self-described dream interpreters to come up with this overview. Believe what you will.

FALLING
Supposedly means: Something in your life isn't going the way it's supposed to; you're not in control.

BEING CHASED
Supposedly means: Your subconscious is trying to get you to face a problem you've been avoiding.

FLYING
Supposedly means: There's an out-of-control issue in your life—something you need to "let go."

DYING
Supposedly means: You may want to terminate something big in your life (a job, a relationship) and get a fresh start.

PUBLIC NUDITY
Supposedly means: You are subconsciously expressing general anxiety and vulnerability—as, for example, when you're about to start a new job.

DRIVING AN OUT-OF-CONTROL VEHICLE
Supposedly means: A current bad habit may be becoming a long-term problem.

MEETING A CELEBRITY
Supposedly means: You may crave recognition yourself—and who the celebrity is may reveal the talents or qualities you personally value.

"

To sleep, perchance to dream –
ay, there's the rub;

For in that sleep of death,
what dreams may come

When we have shuffled off
this mortal coil...

Hamlet

Why Do We Remember Some Dreams But Not Others?

Dreams are wandering and continuous, and don't have logical endings like composed fiction. But the REM stage where virtually all dreaming occurs does have a natural end, and when we wake up before the REM stage concludes, we remember whatever we were recently dreaming about. This can happen when an alarm clock or child wakes you up too soon, or when the dreaming itself snaps you into consciousness: The more emotionally arousing it is, and the more it involves multiple areas of the brain, the more likely it will wake you up—and give you a shot at remembering it. Also, logically well-connected dreams are easier to remember than random and disjointed dreams. If you want to retain your dreams better, it's as simple as telling yourself to do it. "Anything that captures our attention immediately after waking interferes with dream recall, so just as you are falling asleep, keep reminding yourself that you want to remember your dreams," says Deirdre Barrett, author of "The Committee of Sleep," in *Scientific American*.

There's something in humankind that is desperate for our dreams to be meaningful, despite a general rational understanding that dreaming is simply our brain reshuffling what it already knows. For example, when asked what things would make them rethink a decision to fly on an airplane, people rated a dream about a crash as more likely to scare them off than an actual government warning about a potential crash. Studies have shown that we tend to believe our dreams when they support our pre-existing ideas and dismiss them as silly nonsense when they don't—an effect known as **confirmation bias**. Researchers asked subjects to remember a dream in detail and to include their feelings about the dream; they discovered that people found a pleasant dream about somebody they liked more meaningful than a pleasant dream involving someone they didn't like.

In other words, even though we know our dreams are essentially randomly generated, we believe they have meaning—but only when they are in line with what we already believe.

But just because dreams are random doesn't

mean they aren't helpful. Something about the random-shuffle nature of dreaming brings thoughts together in unexpected ways that sometimes produce breakthroughs out here in the real world. Jack Nicklaus fixed his golf swing after seeing himself with a different grip in a dream; it was while dreaming that Einstein conceived the theory of relativity, and René Descartes the scientific method, and Mendeleev the periodic table, and Niels Bohr the structure of the atom, and James Watson the double-helix structure of DNA.

What Is Daydreaming?

Our neurons keep firing even when we're idle or asleep, though at a dramatically lower level. Science calls this a **"resting state" network**, and it's the stage on which daydreaming plays. Daydreaming is a willful shifting away from what you're doing to something you decidedly *aren't* doing—picturing yourself on a beach, for example, instead of sitting in your office cubicle putting off work. Interestingly, the farther out of context you daydream, the tougher it is to remember whatever you were originally doing. When the boss walks back in and interrupts your reverie, daydreaming about an overseas vacation is harder to recover from than dreaming about a weekend at home, and daydreaming about something that happened years ago is harder to recover from than daydreaming about something that happened yesterday. Take note, we daydream less as we get older, possibly because daydreaming is often future-focused.

You can thank color TV for the palette you see in your dreams. Up to 88% of us report dreaming in color today, while only 15% did so before its invention.

THINK ABOUT THIS

Interestingly, recent research suggests that people daydream naturally when they're involved in a mundane task that doesn't use all their working memory—when they have extra processing capacity, so to speak. Daydreaming causes you to shuttle back and forth between your **analytic system** (solving problems) and your empathetic system (managing human relationships), and it can string together useful connections we wouldn't normally make with our conscious minds—producing those "Eureka!" moments.

Over a long, long history of your forebears being attacked and struggling to survive, your brain evolved to be a sort of always-on evaluator, vigilantly aware of threats and opportunity. It's when nothing much is going on that it busies itself constructing "What if?" scenarios, and that's

why people tend to have those light-bulb inspirations in the shower, or while driving to work, rather than in the middle of a busy workday.

Do Animals Dream?

It certainly seems like it. All mammals, some birds, and possibly reptiles exhibit the same rapid eye movement (REM) stage that in humans accompanies a dream state. While in this stage, dogs twitch and bark; and platypuses make shellfish-cracking motions as if they're dreaming of a smorgasbord. In fact, now that we can correlate, through EEGs, specific brain activity with specific behaviors, we can see that the brain activity of animals in REM sleep corresponds to things they do while waking: Rats follow dream mazes, birds rehearse songs.

Many animals display very curious sleeping

habits, which affect the time available for dreaming. In general, smaller animals—like your useless cat—tend to sleep more because of their faster metabolisms, while big land animals like cows and horses may sleep four hours a day or less, and giraffes sleep in increments of five minutes. Even animals at perpetual risk, or who otherwise need to keep moving, have evolved ways to manage sleep: With dolphins, one hemisphere goes to sleep at a time. With birds, REM sleep can last mere seconds, minimizing the risk of danger; some species literally sleep with one eye open, keeping watch for predators. Sharks seem to have their own variations on sleep: Some, for example, hover in a current that keeps oxygen-rich water flowing past their gills.

The science is advancing: We've recently learned to control the apparent dreams of lab animals through sound cues, which may presage a coming age of made-to-order dreams for humans. Buckle up.

Nightmares and Night Terrors

Nightmares, like dreams, occur during REM sleep; when it's a scary or extremely negative dream event that awakens you, we call that a nightmare. (If you don't wake up, it's just called a bad dream.) One interesting treatment for recurring nightmares—including those memory-driven nightmares produced by post-traumatic stress disorder—is called IRT, or Imagery

Rehearsal Therapy. Turns out if chronic nightmare sufferers write down their dream in daytime, then rewrite the script however they like (so the evil clowns could become playful puppies, for example) and finally practice that script, their nightmares decline and sleep improves.

Night terrors are different from nightmares. They occur in the third stage of **non-REM sleep**—the deep, slow-wave sleep in which the brain's metabolic rate slows down. As a result, though you wake up in a panic state and screaming, you're groggy and not completely awake, and you probably won't remember the details afterward. Night terrors are most common among kids 5 to 7 years old, with boys more likely to suffer than girls, but some start before the age of 3, and 1–2% of the adult population suffers from them. Kids who experience them are more likely to sleepwalk, too. Night terrors tend to happen in the first half of the night, and one reportedly successful strategy is timing when they happen and waking the child up gently just before the expected episode.

Lucid Dreaming

Most of us are passive characters in our dreams and nightmares, a point of view wan-

If you have a child who sleepwalks, try waking her up about 45 minutes after she goes to sleep to break the cycle, recommends Carlos Schenck of Minnesota's Regional Sleep Disorders Center.

SCREEN TEST

Did *Inception* get lucid dreaming right?

It wouldn't be surprising if it did. Not only is *Inception* writer/director Christopher Nolan a lucid dreamer, but star Leo DiCaprio reportedly had lucid dreams before filming. The concept that "dream time" moves at a different pace than time in the real world accords with the way lucid dreamers describe it, so that's a good start. But other details don't ring true: The foundational dream-within-a-dream concept isn't how lucid dreamers describe the experience (they'd just play out as consecutive dreams), and two or more people sharing one dream is still the stuff of fantasy. There's more: "In my history of having more than 1,000 lucid dreams, I've never been involved in a gun battle or a car chase," says author Robert Waggoner, author of multiple lucid dreaming books. Though he does have personal experience with lucid dreaming, director Nolan seems to have used it merely as a storytelling device, probably one reason the phrase "lucid dreaming" appears nowhere in the script or the movie.

UNREAL
The average person will have more than 100,000 dreams in their lifetime.

SIX FAMOUS LUCID DREAMERS

Lucid dreaming, under different names, has been around for a long time. Here are some of its more famous practitioners:

- **Richard Linklater**, writer/director of *Dazed and Confused* and *Waking Life*
- **Christopher Nolan**, writer/director of *Inception*
- **James Cameron**, director of *Titanic* and *Avatar*, who said creating *Avatar* was all about trying to create a lucid dream experience
- **Richard Feynman**, genius theoretical physicist
- **The Wachowskis**, creators of the *Matrix* series (which has been described as a guide to lucid dreaming)

dering from one scene to the next. But sometimes the "dream you" understands that he is a person who's dreaming, and in some cases even has some free will and can decide whether to follow the strange rabbit in a waistcoat down that rabbit hole. This is called a **lucid dream**, and learning to control this process, if its proponents are to be believed, can turn sleep from a necessary evil into an opportunity, where you can control your own dreams and guide yourself to new experiences, and even get better at certain activities through virtual practicing. "Once you know that you're dreaming as you're dreaming, you gain access to the most powerful **virtual reality** generator in existence: the mind," says Charlie Morley, author of *Lucid Dreaming*, on FastCompany.com.

Sure, it sounds like something out of a movie. Lucid dreaming is the subject of *Inception*, among other films, and it's how the teenagers—spoiler alert—*finally* beat Freddy Krueger in *A Nightmare on Elm Street*. But scientists say lucid dreaming is no delusion. They've measured higher alpha waves in dreamers

who claim to be lucid dreaming, and experienced lucid dreamers can actually tell researchers when they're entering the lucid state with physical cues like switching their gaze from left to right, *while they're asleep.* When lucid dreamers were asked to dream about engaging in physical exercise, they claimed to be doing so—and their heart rates went up accordingly.

There's a marked similarity between the lucid dream state and the wakeful state, as if your brain really thinks you're awake and directing it. This leads to a fairly shocking result: When you practice something while lucid dreaming (say painting or playing piano or even running a marathon) *it can yield the same benefits as if you'd practiced in the real world.* Obviously, dreaming can't teach you a skill you don't know or build muscle while you're curled up on the couch. But in terms of the mind-body control that's so key to many activities, you can, in fact, *dream yourself better.* One lucid dreamer, for example, quit

smoking by only smoking inside lucid dreams—now he gets all of the pleasure, none of the cancer.

How is lucid dreaming done? In short, you need to realize you're in a dream without that realization being so shocking that it wakes you up. So be cool, and follow these steps:

- First, put a dream journal by your bed and capture everything you can remember. It has to be nearby, because real-world distractions quickly derail you and make you forget dream detail.

- Repeat a mantra to yourself, as you fall asleep: Tell yourself that you're going to remember your dreams.

- Work the snooze button: People who use it have more lucid dreams, perhaps because being awake for just a minute or two can drop you back into REM sleep, where dreams happen.

- While dreaming, try to watch for "triggers" that can prove you're in a dream. Look for a clock—they're hard to understand in the dream state. Or try to read and reread a sign or a book—the words may change.

- Spend some time, while awake, reviewing your dreams and imagining yourself acting differently in them.

Et voilà! If you find yourself suddenly conscious that you're in a dream, try not to get too excited; that can break the spell and wake you up. And be advised, if you delve into this, that some frequent lucid dreamers have trouble on the flipside: They can be unaware they're not dreaming when they're in the real world. Now, *that's* the stuff of nightmares.

What Do Conscious, Unconscious, and Subconscious Mean?

The terms aren't always used precisely, but conscious generally means the state of being aware of what's going on around you; unconscious is a state where you're totally unaware and not thinking in any meaningful sense. The subconscious sits between those two simple end-states and is much more interesting.

Subconsciousness could be described as the state in which your brain receives inputs you're unaware of. The vast majority of thought happens at the subconscious level—there's much more going on in the world around you than you can afford to pay direct attention to. When there's a song you can't get out of your head, when random thoughts interrupt your exam studies, or when you finally become conscious of a sound that's surely been going on for a long time, you're getting a glimpse of your subconscious mind. Your conscious mind tells you what to focus on from a vast sea of sensory inputs and their references and connections. Your subconscious mind provides the sea.

Breathing is one example of an unconscious, or automatic, activity (unless you interfere by consciously holding your breath). The reflex action that pulls your hand off the stove before you feel the heat might be considered a subconscious activity: Your mind and body are deciding and acting quickly based on inputs before you are totally aware of what's going on.

Remarkably, scientists have recently found the "on/off" switch for consciousness. The culprit is the **claustrum**, a sheet-like organ deep in the middle of your brain that seems to coordinate the high-level inputs from other brain areas into one coherent story, rather than a chaotic flurry. By stimulating the claustrum with high-frequency electrical impulses, scientists have demonstrated they can literally turn off (and on) consciousness, such that the subject immediately glazes over, then recovers when the stimulation stops.

Recent research has begun to illuminate the remarkable depths that subconscious thinking can have on our behavior without our knowing it.

For example, one study found that when students ate a crumbly snack in a room scented with a mild cleaning product, they were three times more likely to clean up after themselves than students in a room of neutral scent. In another study, subjects played an investment game more aggressively when there was a slick black briefcase on the table

HOW TO TAKE A
POWER COFFEE NAP

If you've got half an hour and need a nap, caffeine can be your friend.

The neurochemical adenosine makes you feel tired, and caffeine keeps sleepiness at bay by blocking adenosine's receptors. But it takes 20 minutes or so for the caffeine to kick in—long enough for a quick nap. By sleeping while you're processing that double macchiato, your body's producing less adenosine for the caffeine to have to deal with, providing a more powerful pick-me-up.

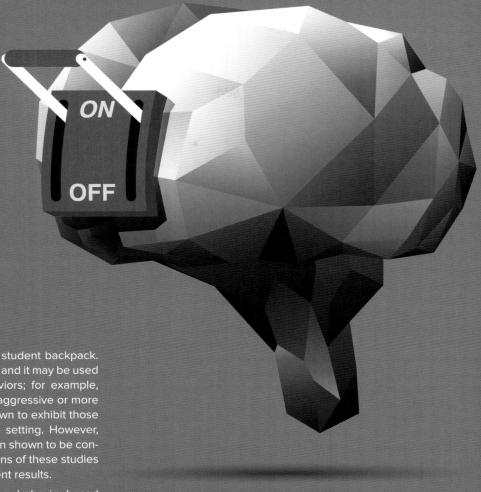

across from them versus a student backpack. This effect is called priming, and it may be used to change individual behaviors; for example, people primed to be more aggressive or more cooperative have been shown to exhibit those symptoms later in a group setting. However, priming studies haven't been shown to be conclusively reliable—replications of these studies have, at times, found different results.

How can the brain adjust your behavior based on inputs you're unaware of? The theory is that some decision-making is handled by the more primal, emotional parts of the brain that lie closer to the brain stem and far from the more recently evolved, and more rational and disciplined, cortex. Fight-or-flight decisions, for example, seem to be made at this primal level, before the information makes its way on up to the higher-function areas.

Advertisers and marketers have long taken advantage of your weakness for subconscious inputs, with a whole raft of things designed to move you in their direction without your knowing it, like putting their neutral messaging in the voices of actors you love or placing flowers prominently at the front of the grocery store to send a subconscious message that the food is fresh. By manipulating your mood through music, scents, and other artificial inputs, they hope to move you incrementally to buy their products before you even know what you're

doing. It works...and it's incredibly hard to resist.

Everything we've covered so far—the structures of the brain and how they interact, the chemicals that deliver rewards and induce emotions, the sensory inputs and how they're disentangled, how thoughts and memory happen, and how sleep and dreams work—hopefully present a good general idea of how the brain functions when everything's working. But as with any fantastically complex system, in a wide world of problems and disasters, things can and do go wrong. From injuries to strokes, from **phobias** to manias, from hereditary conditions like Alzheimer's disease to conditions like post-traumatic stress disorder, the "broken" brain—and whether it's really "broken" at all—is the subject of the next chapter. 🧠

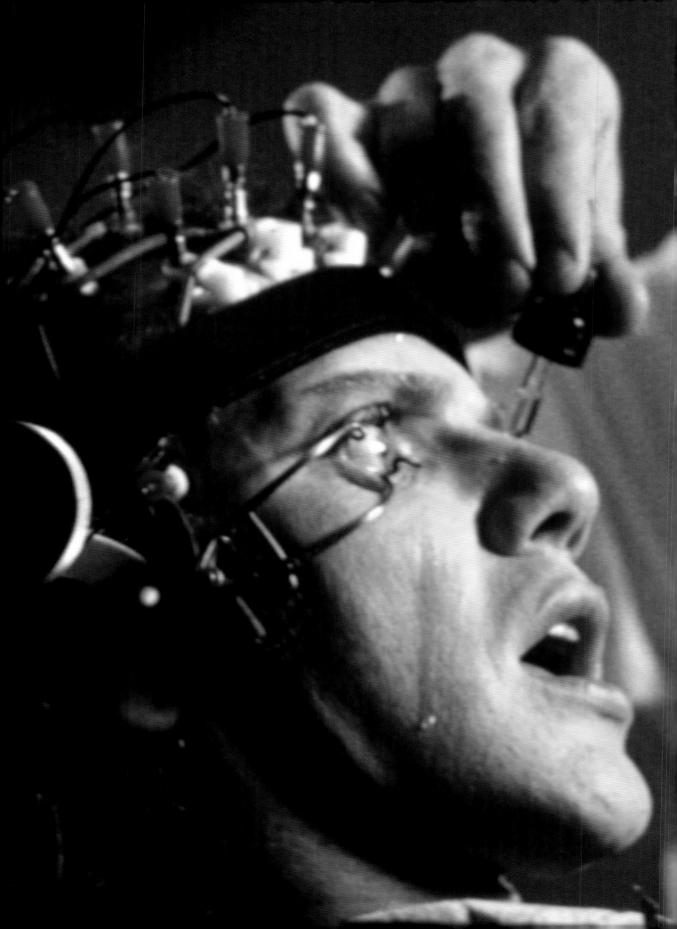

CHAPTER 5

DISEASE, DISORDER, AND THE "BROKEN" BRAIN

There's been an unspoken assumption in everything we've explored so far: a smoothly functional brain, humming along happily. But what happens when things go wrong—disastrously wrong?

The man in the dusty 19th century daguerreotype is handsome. Clean-shaven, hair parted to his right, unsmiling as per the apparent protocol of the age. But this isn't an ordinary, awkward portrait. The first clue—the gaze of one eye is piercing, but the other eye is squeezed shut. The second clue—the long, pointed spike the man is clutching in both hands. The owners of the photo, vintage photograph collectors Jack and Beverly Wilgus, wrongly assumed the spike was a harpoon, and the man an injured whaler. But after posting the picture online several years ago, the Wilguses came to realize that they owned what may be the only existing photograph of Phineas Gage, the most famous medical case study in the history of brain trauma.

Gage was a railroad worker, not a whaler, and the spike in his hands is a tamping iron, used to pound sand and explosive powder into holes back in the days before class action lawsuits. In 1848, the 25-year-old was pounding away with this tool when the powder detonated, and the explosion fired the 13-pound, 43-inch long pipe up through his face and out of his skull, re-

MIND BENDERS

25

Percent of U.S. adults who have at least one mental illness, according to a study cited by the Centers for Disease Control and Prevention (CDC)

18

Percent of all adult suicides in the U.S. that are veterans

9

Percent of all violent crimes committed by those with psychiatric disorders

The #1 cause of disability worldwide is depression.

moving part of his brain and severely rupturing the optic nerve in his left eye. The fact that Gage survived such a trauma in the "just add leeches" 1800s is miraculous enough, but what doctors discovered afterward made Gage a medical miracle. Because although he survived, Gage underwent a radical personality change. Formerly a self-controlled character, he became erratic, showing "little deference for his fellows," and had gained the new habit of spewing "the grossest profanity." Yep: He became *that guy*. Friends complained that he was "no longer Gage," and presumably took him off the active party circuit.

Strange.

Gage's beam-through-the-head mishap was inadvertently reenacted by another man 130 years later with very different results. In 1978, at the Institute for High Energy Physics in a small town just south of Moscow called Protvino, a 36-year-old researcher named Anatoli Bugorski was working on the U-70 Synchotron, then the most powerful particle accelerator in the world. The machine appeared to malfunction, and Anatoli unwisely leaned in to investigate—and that's when the machine sent a proton beam in through the back of his skull and out the other side, just to the left of his nose.

Anatoli later described it as a flash "brighter than a thousand suns," though mysteriously painless. Not only did the incident not kill Anatoli, despite his exposure to radiation, he survived with partial facial paralysis, loss of hearing in his left ear—*and* retained full mental capacity. Though he would struggle with lifelong fatigue, he continued working, completed his PhD, and today lives a relatively normal life.

The brain is astonishingly resilient. It has to be, since it's under a more or less constant bombardment from various threats. The skull offers a good deal of protection against blunt force trauma, and the blood-brain barrier does yeoman's work keeping out most disease and invaders. But injuries and illnesses happen, and microscopic monsters crash the gate, and sometimes, the trouble arises from within.

And brain trouble is scary at a profound level because it threatens identity itself. Breaking your arm can be debilitating, and confronting bypass surgery terrifying, but there's little risk of either incident fundamentally changing who you are as a person. But whether we're discussing sports-related **concussions**, post-trau-

matic stress, genetic abnormalities, or paralyzing behavior and control issues, the threat is existential—a risk that the condition, and/or its treatment, will fundamentally alter who you are.

What's a body to do when there's trouble at the top?

The Long and Terrible History of Our Attempts to Fix Our Brains

Since ancient times, humankind's collective fear of losing our minds has led us to demonize and ostracize those perceived to have disorders of the mind. Worse, their suffering, in all too many cases, paled beside the pain we inflicted with our boneheaded attempts to help.

THE FIX: TREPANATION
(Prehistory - Renaissance)

For all too much of human history, talking smack to your tribe's shaman or the village elders was *de facto* evidence of a demon living inside your head. Luckily, there was a cure—unluckily, it involved drilling a little birdhouse hole in your skull to release the demon inside. **Trepanation** was quite common, to judge from the fossil record, and drilling or scraping the bone away to reveal the brain itself is what passed for psychotherapy back when a sufferer might have been deemed a witch or worse. Some unenlightened *Homo sapiens* still practice voluntary trepanation to this day, on the advice of dubious proponents who claim it of-

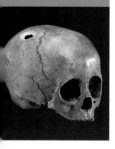

5-10% of all Stone Age skulls found have trepanation holes in them.

We used to call insane people "lunatics" because **we thought they were under the influence of the moon** (Latin, *luna*).

fers benefits like increasing blood flow to the brain. Science is, shall we say, skeptical.

THE FIX: INSANE ASYLUMS *(1600s-1900s)*

As the Middle Ages gave way to an era of increasingly larger urban environments, "madmen" began to be moved out of homes and communities and into state-run institutions like London's Bethlem Royal Hospital (whence we get "bedlam," a word for frenzied chaos). Unfortunately, places like this—which focused more on removing the disturbed from public view than trying to treat them—were often designed more for the benefit of society than for the patients. Essentially, these were convenient places to dispose of—often permanently—socially disruptive individuals of any kind, including opinionated women.

THE FIX: PSYCHOANALYSIS *(1890s-today)*

Austrian neurologist and notorious cigar-noticer Sigmund Freud conceived the basics of **psychoanalysis** while treating anxious patients, hypothesizing that some of that anxiety could result from repressed memories that were not consciously remembered by the sufferer but which could be drawn out through focused discussion. His views on dreams as windows into your unconscious wishes and anxieties, and sexuality as a progression through distinct oral, anal, and phallic phases, haven't aged well. And although psychoanalysis has long been in decline as a clinical practice, Freud's notions of the **id** (your unconscious, animal, instinctual desire), **ego** (your rational decision-making intelligence) and **superego** (your moral, society-driven "better self") continue to influence medicine and culture even today, more than a century later.

THE FIX: INSULIN-INDUCED COMAS *(1920s-50s)*

When physician Manfred Sakel accidentally gave a morphine-addicted diabetic too much insulin in 1927, she went into a coma, but revived and proclaimed her morphine cravings gone. Further experimentation and more examples of dramatically improved mental states in patients convinced the ambitious Sakel that he'd stumbled on the cure for everything from addiction to **schizophrenia**. Successes in Europe and America brought Sakel many years of fame, but the efficacy-to-risk ratio started being questioned in the '40s, and by the '50s, the practice of inducing comas declined sharply in popularity, as more effective and less dangerous pharmacological interventions began arriving on the scene.

THE FIX: INDUCED CONVULSIONS *(1930s)*

At around the same time, a Hungarian doctor named Ladislas von Meduna found that inducing convulsions with the drug metrazol seemed to help patients with schizophrenia. There was one tiny side effect: The convulsions were so severe that they caused spine fractures in two out of five patients. While they found a way to neutralize this effect by pairing the metrazol with the poison curare—were there NO lawyers in the '30s?—induced convulsion therapy gradually lost favor in the face of more promising treatments.

THE FIX: ELECTROSHOCK TREATMENTS/ ELECTROCONVULSIVE THERAPY (ECT) *(1930s-today)*

Inspired by watching doomed pigs get electroshock as a form of anesthesia before their butchering, Italian neurologist Ugo Cerletti decided to try it on human schizophrenics. **Electroconvulsive therapy**, or (ECT), in its benevolent modern form, worked wonders on acute-onset schizophrenia, and later on depression. ECT had the added benefit that it provoked **retrograde amnesia**, so the subjects forgot what had just happened. ECT quickly replaced the more problematic seizure-inducers above, and though its reputation took a major hit with 1975's *One Flew Over the Cuckoo's Nest*, ECT remains one of the most effective treatments for severe depression,

Five Famous People Who Received **Electroshock Treatments/ECT**

- **Sylvia Plath** received ECT for severe depression. *"A great jolt drubbed me till I thought my bones would break and the sap fly out of me like a split plant."*
- **Dick Cavett** called ECT a "magic wand" for his depression.
- **Ernest Hemingway** was given ECT to treat depression, but blamed it for a career-ending loss of memory, and committed suicide shortly afterward.
- **Kitty Dukakis** says ECT helped relieve her alcoholic depression.
- **Lou Reed** was given ECT, at age 17, reportedly to "cure" his bisexuality. (Needless to say, it didn't work.)

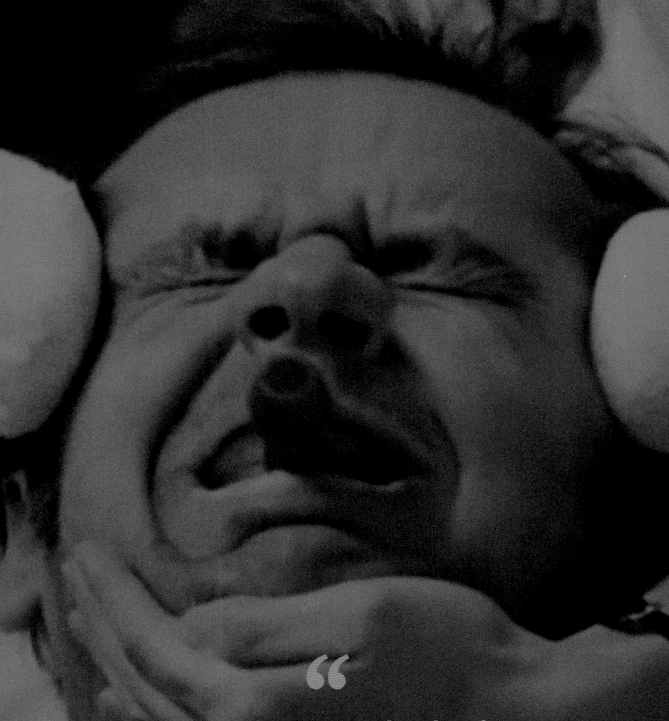

"

What is the sense of ruining my head and erasing my memory, which is my capital, and putting me out of business? It was a brilliant cure but we lost the patient.

–**Ernest Hemingway**, who'd received electroshock therapy that ruined his ability to write; he would commit suicide shortly thereafter.

Three Great Inventions We Got From Wrong-Headed Treatments

THE VIBRATOR Invented by Joseph Mortimer Granville as a muscle relaxer for men in the 1880s, it was quickly adopted by doctors of the day as a treatment for women, for the then-commonly diagnosed condition called "hysteria." Specifically, the electrical device relieved doctors from the cramp-inducing work of manually stimulating patients' clitorises—the common treatment of hysteria. The vibrator was born—and the original, awesomely, was known as "Granville's Hammer."

GRAHAM CRACKERS The original bland, unsweetened formula was developed by Presbyterian minister Sylvester Graham as an alternative to the sugary crackers he believed were leading children to dangerous "self-abuse" (e.g., masturbation). Nabisco turned the tables, sweetening them into the s'more-supporter we know and love today.

CORN FLAKES Similarly hopeful that bland food would curb that ol' devil masturbation, Dr. John Harvey Kellogg created corn flakes for patients at his sanitarium. His brother, W.K., saw a wider market for the product and added sugar to the recipe, starting a blood feud that would last decades. Fun fact: Dr. Kellogg also recommended that girls' troublesome sexual desire could be curbed by burning their ladyparts with carbolic acid. So there's that.

and is still performed on more than 100,000 people a year.

THE FIX: LOBOTOMIES *(1940s)*
The idea that you could literally cut mental illness out of the brain was the brainchild of a Portuguese doctor named Egas Moniz. Moniz observed that removing the frontal lobe from monkeys stopped them from throwing their own feces, and he began experimenting with prefrontal **lobotomy** on humans, hoping to cure schizophrenics by surgically fusing nerves in the brain that were thought to be the source of the problem. The results of this procedure were so promising that Moniz was awarded a 1949 Nobel Prize, and presided over thousands of the procedures. And so did others: American doctor Walter Freeman drove a "lobotomobile" around the country treating everyone from schizophrenics to depressed housewives; he sometimes inserted a pick into the eye socket to pierce the thin bone at the back to sever connections in the brain. But while often effective, the procedure could also induce a kind of permanent lethargy, and sometimes epilepsy and death, and by the '50s, lobotomies were fast falling out of favor.

THE FIX: ANTIDEPRESSANTS *(1950s-today)*
Scientists researching cures for schizophrenia at the Münsterlingen asylum in Switzerland stumbled upon a drug that gave patients a blast of euphoria. It didn't help their target schizophrenics, but the drug, dubbed Tofranil, was seen to help with depression, and hence an antidepressant market was born. This effective, noninvasive treatment option was an instant hit, and successors like Prozac, Zoloft, and Paxil built on Tofranil's early success. Today around 10% of Americans take one or more antidepressants, and their effectiveness in treating severe depression is recognized fairly universally. However, the pharmaceutical industry has been accused of over-marketing the drug to the much larger body of people with mild to moderate depression, for which antidepressants' effects are minimal to nonexistent, according to a Harvard Medical School study.

Brain Conditions: Everything You Need to Know

A quick portfolio of pain: Here's what happens with—and what we can do about—various disorders of the brain and mind.

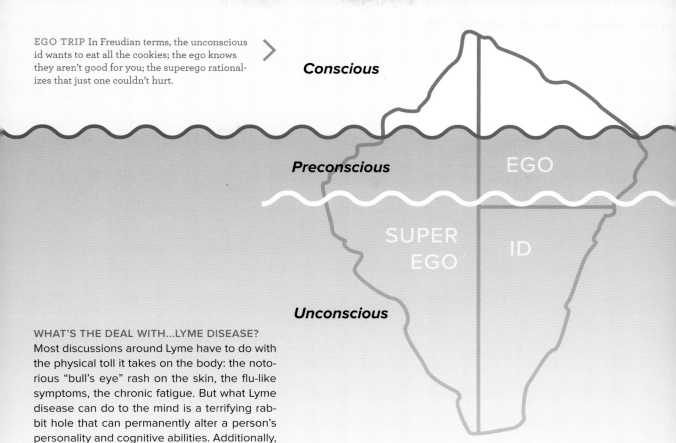

EGO TRIP In Freudian terms, the unconscious id wants to eat all the cookies; the ego knows they aren't good for you; the superego rationalizes that just one couldn't hurt.

Conscious

Preconscious

EGO

SUPER EGO

ID

Unconscious

WHAT'S THE DEAL WITH...LYME DISEASE?

Most discussions around Lyme have to do with the physical toll it takes on the body: the notorious "bull's eye" rash on the skin, the flu-like symptoms, the chronic fatigue. But what Lyme disease can do to the mind is a terrifying rabbit hole that can permanently alter a person's personality and cognitive abilities. Additionally, it can apparently be passed to unborn children. In a story published on ScientificAmerican.com, a 26-year-old woman named Justine Donnelly who'd suffered her whole life with poor memory and anxiety eventually tested positive for Lyme, although there were no signs that Donnelly had been bitten by a disease-bearing tick. Instead, doctors believe that Donnelly's mother had contracted the disease three decades earlier and passed it to Justine in the womb. What Donnelly thought were innate characteristics were actually the results of a disease fundamentally altering her brain chemistry before she was born.

An extreme case? Sure. But in addition to feeling physically exhausted, feverish and ill, Lyme has been shown to attack the nervous system with a vengeance, leading to memory loss, erratic mood changes, depression, panic attacks, hallucinations, and dementia. Archeologists have unearthed 15-million-year-old ticks preserved in amber that were shown to be carrying the disease, and in 1991, other archeologists found the mummified remains of a 5,300 year-old man, dubbed Ötzi the Iceman, that was also infected. Lyme disease has been with us a long time, and it isn't going away anytime soon.

WHAT'S THE DEAL WITH...ALZHEIMER'S DISEASE?

Perhaps no mental condition inspires more general dread than the long, slow mental decline of Alzheimer's disease, a progressive brain disorder that lays waste to memory and cognitive ability. Sufferers slowly lose memory, the ability to walk or speak or take care of themselves, and even, in the later stages, mission-critical bodily functions like coughing and breathing. The CDC lists Alzheimer's as the sixth leading cause of death in the United States; patients survive about seven years, on average, from onset. Alzheimer's occurs when deposits of proteins in the brain form blockages called "plaques" or "tangles" that start disrupting the function and connections of healthy neurons, although it's unclear whether these are a *cause* or *symptom* or correlate. There's no cure, but symptoms can be treated, and promising diagnostic interventions include scans to detect plaque tangles even before memory loss starts.

The lack of clarity around Alzheimer's has inspired a well of urban legends to rival the de-

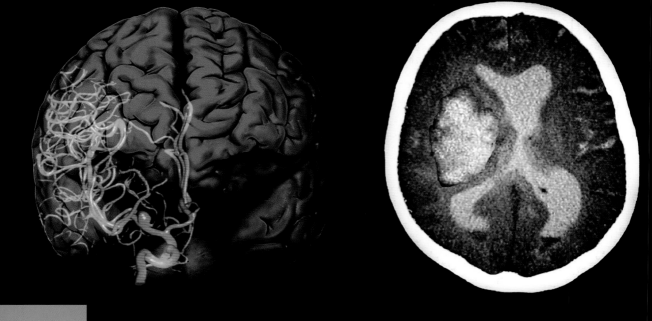

bunked myth of children's vaccinations causing autism. In the '60s and '70s, it was believed that drinking out of aluminum cans, cooking with aluminum pans, and even using deodorant (which contained trace amounts of aluminum) could all lead to Alzheimer's. Another faction held that the compounds used by dentists to fill in cavities, which contained mercury, silver, and tin, were setting everyone on the path to mental degeneration. Artificial sweeteners like aspartame, found in products like NutraSweet and Equal, were also suspect. Even if you somehow avoided those, it was believed that flu shots also contained the perfect Alzheimer's cocktail. These notions have been pretty conclusively debunked, but their lingering popularity speaks to the dread that surrounds this heartbreaking, personality-eroding disease.

WHAT'S THE DEAL WITH...STROKES AND ANEURYSMS?

If a heart attack is what occurs when there's a loss of blood flow to the heart, you might think of a stroke as a "brain attack." An **aneurysm** is a balloon-like bulging in the the wall of an artery. A brain aneurisym can press on areas of the brain and cause symptoms like blurry vision or neck pain, otherwise it can go undetect-

ed. But if it ruptures, the subsequent internal hemorrhaging and lack of blood flow to the original destination can cause a stroke. Where the stroke occurs matters: An attack in the right hemisphere, which heads up cognition, emotions, and spatial orientation, can result in diminished attention span, short term memory loss, and loss of the ability to process visual and verbal information. An attack in the left hemisphere, which is primarily in charge of talking and expressive language skills, can lead to severe problems forming and expressing words, a condition called **aphasia**. (You might say a right-hemisphere stroke can limit your ability to process inputs; a left-hemisphere stroke can limit your ability to produce outputs.) The severe loss of functionality a stroke can deliver, and the lifestyle changes it can necessitate, makes it among the most dire of brain injuries; a fundamental blow to the sufferer's core self. "I felt my spirit surrender, and in that moment I knew that I was no longer the choreographer of my life," said neuroanatomist Jill Bolte Taylor, describing her own stroke in her bestselling book and popular TED talk, "My Stroke of Insight."

WHAT'S THE DEAL WITH...POST-TRAUMATIC STRESS DISORDER?

Post-Traumatic Stress Disorder (PTSD) is a

"No great mind has ever existed without a touch of **madness**."

–*Aristotle*

debilitating condition resulting from physical changes in the brain following severe and/or prolonged exposure to damaging, horrific environmental stimuli. As with most forms of brain damage or impairment, the specific type of abuse or trauma can have vastly different effects on the brain. But it's often found in war veterans, victims of domestic and sexual abuse, and other survivors whose brains show real physical damage from the terrible, emotionally charged incidents they've endured.

The condition is complex, involving damage to the fear-generating amygdala, the memory-managing hippocampus, and the fear-suppressing, context-establishing, cerebral cortex. Rather than merely *remembering* something horrible, PTSD sufferers seem to be reliving it, over and over. It can affect their ability to interpret environmental context—if your PTSD stemmed from an incident where you were shot in a parking garage, say, being in a similar environment could trigger a stress response, as the brain struggles to decide if this is a new and completely unrelated parking garage or if you are reliving the traumatic past. For war veterans, the boom of an explosion in a violent movie can kick-start the body's adrenaline, as if the brain believes, at some primal level, it is back in the thick of battle.

PTSD, under other names, like "shell shock," as it was called during WWI's horrific trench warfare, has probably been around for the entire course of human history, but it's only recently become clear just how far-reaching its effects can be. The **somatosensory cortex**—the place where your body is "mapped" inside your brain—can vary in thickness, with a thicker somatosensory cortex roughly corresponding to a healthier individual. In the cases of victims of rape or sexual assault, the cortex thins around the areas mapped to genitalia: The abuse they experienced altered the victims' brains' internal maps of their bodies, leaving actual "emotional scars." Similarly, a thinning in the cortex areas assigned to self-awareness and emotional reg-

TRAUMA
For about 1 out of 10 Americans who suffer a traumatic event—war, a natural disaster, sexual assault—it triggers PTSD.

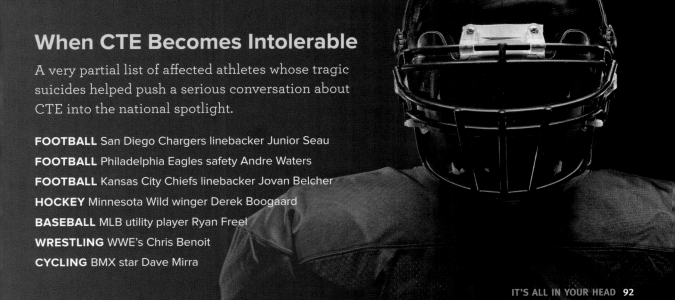

When CTE Becomes Intolerable

A very partial list of affected athletes whose tragic suicides helped push a serious conversation about CTE into the national spotlight.

FOOTBALL San Diego Chargers linebacker Junior Seau
FOOTBALL Philadelphia Eagles safety Andre Waters
FOOTBALL Kansas City Chiefs linebacker Jovan Belcher
HOCKEY Minnesota Wild winger Derek Boogaard
BASEBALL MLB utility player Ryan Freel
WRESTLING WWE's Chris Benoit
CYCLING BMX star Dave Mirra

ulation is seen in patients who've been emotionally abused throughout their lives.

WHAT'S THE DEAL WITH...CONCUSSIONS?

A concussion occurs when a blow or trauma of some kind bounces your brain against the inside of your skull, causing damage. (It could happen when you're sacked by a linebacker or rise too fast on the train and bang your noodle hard on the stupid *#$!%@! luggage rack.) Symptoms include headaches, balance and memory problems, and reduced concentration, even for weeks or months after the initial trauma. Traditionally, concussions were dismissed as no big deal, but that turns out to have been wishful thinking. A 20-year study of over 200,000 people showed that the risk of suicide rises threefold after just one mild concussion, with the suicides tending to happen around six years after the initial trauma, on average. Clearly, this is serious business, and the world is starting at long last to take notice.

Athletes, particularly in contact sports, are of course at high risk for concussions, and though they're often declared "recovered" and returned to play in a week or two, the brain damage is still visible six months later. The NHL records about 80 concussions per season, with sufferers taken out of play for a median of six days, and more than a third of water polo players—who wear light ear-protective caps, not helmets—report having had at least one concussion. Helmets generally help prevent skull injuries like fractures, but they don't do much for concussions, where the injury is the brain colliding with the inside of the skull.

Long-term damage from repetitive small brain injuries that stretch and sometimes snap axons, the long connectors between different parts of the brain, is known as **CTE (Chronic Traumatic Encephalopathy)**, as the NFL and other leagues have recently and somewhat reluctantly begun to recognize officially. CTE can't currently be diagnosed until after death; but it's believed to be so widespread as to be almost universal among athletes in contact sports. In fact, a 2017 Boston University study of deceased NFL players found CTE in 110 out of 111 (99%)

tested. And it isn't just professional sports: CTE has been found in athletes as young as 17; in one study, more than 91% of former college football players—who'd never gone pro—were revealed to have CTE. Recently, athletes like NASCAR's Dale Earnhardt Jr. and soccer's Mia Hamm have promised to donate their heads to science to help us figure this out.

WHAT'S THE DEAL WITH...ANXIETY?

It's natural and important to feel worried in certain situations, but persistent feelings of fear, apprehension, and nervousness that linger across situations without real triggers indicate an anxiety disorder. Anxiety disorders, which range from **generalized anxiety disorder** to **panic disorder** to phobias, are another one of the most prevalent illnesses worldwide: About a third of the global population will experience an anxiety disorder at some point in their lives. Anxiety often coexists with depression, and together they cost the global economy $1 trillion annually, according to the World Health Organization.

The anxious brain, flooded with the fight-or-flight hormones cortisol and **norepinephrine**, is on high alert for potential threats—real, imagined, or perceived—and overreacts to neutral and non-threatening stimuli, a behavioral phenomenon called **overgeneralization**. The main culprits in the brain that are thought to promote anxiety are the amygdala, hippocampus, and the lateral septum, a midbrain structure attached to the hippocampus that communicates with the hormone-driving hypothalamus. Anxiety can be difficult to treat—anxiolytic medications only help half of all patients who try them—but cognitive behavioral therapy has been found to help the majority.

WHAT'S THE DEAL WITH...DEPRESSION?

Depression is a very common ailment that's different from sadness and grief, which are rational, temporary responses to negative experiences. Depression lasts at least two weeks, by clinical definition, and is characterized by a much more general negativity: a loss of pleasure in once loved activities, disruption in eating and sleeping, feelings of worthlessness and despair. Unlike grief, it's typically unmediated by relieving moments or days of positivity. The depressed brain shows overactive amygdalae, decreased activity in the ventromedial prefrontal cortex (which normally regulates activity in the amygdalae), sensory

PROTECTING YOUR
PRECIOUS CARGO

Sometimes it's good to be a birdbrain.

HOW DOES THE Q-COLLAR WORK?

1 Invented by Q30 Innovations, the horseshoe-shaped plastic unit rests around the back of the neck, like a travel pillow.

2 The unit compresses the jugular vein gently—similarly to the pressure applied by a necktie, according to the creators—slightly slowing blood flow out of the brain.

3 The extra teaspoon of blood retained between brain and skull leaves less room for the brain to move around, resulting in a snugger fit within the skull.

4 The Q-Collar draws its inspiration from the anatomy of the woodpecker. Its long tongue and the supporting bone and muscle structures surround the jugular vein, cushioning the brain as the bird chisels away.

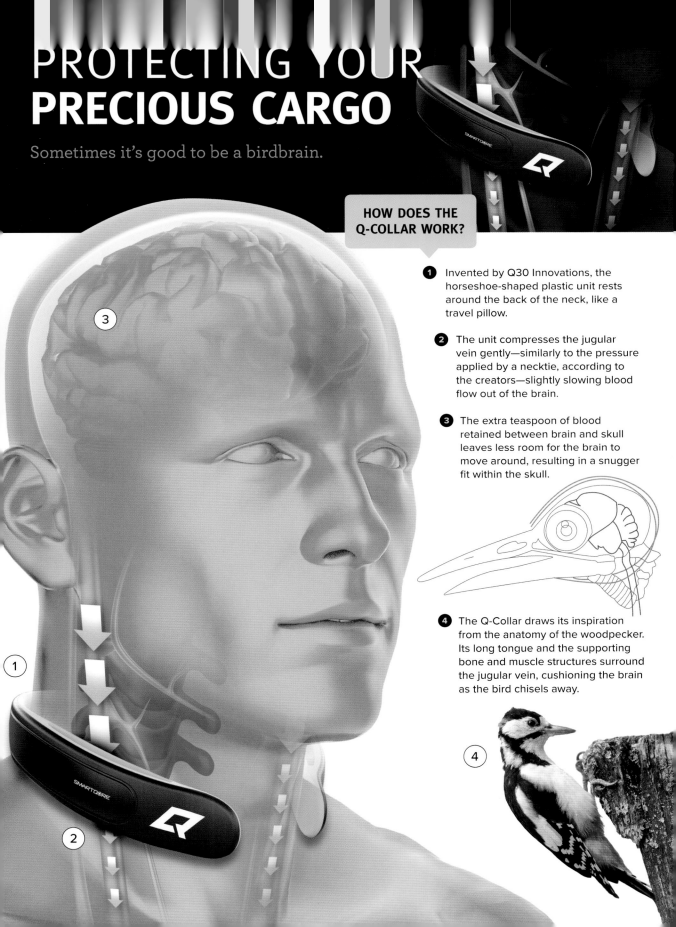

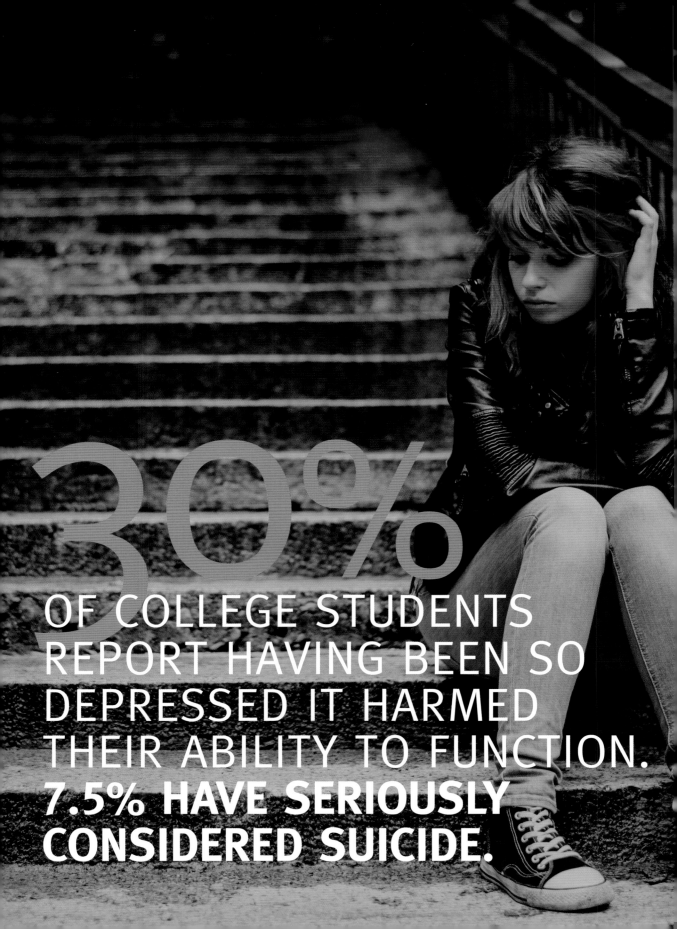

30% OF COLLEGE STUDENTS REPORT HAVING BEEN SO DEPRESSED IT HARMED THEIR ABILITY TO FUNCTION. **7.5% HAVE SERIOUSLY CONSIDERED SUICIDE.**

processing issues in the thalamus, and a reduction in **dopamine** sensitivity in the nucleus accumbens. Recurrent depressive episodes deepen the damage to the brain, eventually shrinking the hippocampus.

Depression is one of the world's most prevalent illnesses: It affects about 5% of people worldwide at any given point in time, with women more likely to experience it than men.

WHAT'S THE DEAL WITH...OBSESSIVE COMPULSIVE DISORDER?
Obsessive Compulsive Disorder (OCD) is an oft-misunderstood anxiety disorder that's far more rare and debilitating than popular belief would have it. The two key diagnostic elements are an obsession—intrusive thoughts, impulses, and images—paired with a compulsion—an uncontrollable and counterproductive urge that must be satisfied at all costs in order to soothe the triggered anxiety. While a definitive cause hasn't been identified, science points to low serotonin levels and communication errors between the social orbitofrontal cortex, the emotional and motivational **cingulate gyrus**, and the caudate nucleus, part of the **basal ganglia** that controls voluntary movement.

Although many claim to have OCD—say, your manager, who goes through a bottle of hand soap every workweek—they're more likely to be demonstrating obsessive and compulsive *tendencies* that are completely normal. OCD only affects 1% of the U.S. population. Those diagnosed are unable to control their obsessions or compulsions no matter how self-aware they are; when triggered, their brains hijack their bodies into action against their will. This lack of control over their own thoughts and actions causes significant distress and problems at work, school, and in social life. Even OCD patients who exhibit hygienic compulsions like excessive hand washing do it not for the sake of cleanliness, but to help cope with a fundamental fear of exposure to potential illness or contamination. As for your fastidious manager? She's probably just a run-of-the-mill germophobe.

WHAT'S THE DEAL WITH...SCHIZOPHRENIA?
Schizophrenia is a chronic and severe mental disorder characterized by disorganized thinking and behavior, and gives rise to a host of ailments, like delusions and hallucinations that cause a sufferer to be generally "out of touch" with reality. But it can also cause a host of milder symptoms, like reduced feelings of pleasure, an inability to feel or express emotion, and troubles with focus and working memory. It's often confused with **dissociative identity disorder** (a.k.a. multiple personality disorder), a rarer condition in which a normally multifaceted human identity "splinters" into multiple personalities living within the same mind/body. Schizophrenia involves imbalanced levels of at least three neurotransmitters: dopamine, serotonin, and glutamate. Some scientists, noting that the brains of schizophrenic patients have fewer connections between neurons, believe the culprit is overactive synaptic pruning: This natural weeding out of unnecessary connections between neurons is very active during adolescence—when schizophrenia symptoms typically appear—as brain regions hone more specialized functions and reduce sensory overstimulation.

WHAT'S THE DEAL WITH...AUTISM?
If schizophrenia is the result of not enough connections within the brain—one part isn't talking to another, and the brain's behavior police are asleep behind the wheel—**Autism Spectrum Disorder (ASD)** may be the flipside of that coin, the result of too many connections within the brain. A Columbia University study compared the brains of children and adolescents with and without autism: In neurologically typical kids, synaptic pruning during the explosive growth of early adolescence typically weeds away about 40% of the connections made since early childhood, but in autistic patients, synaptic pruning had only reduced 16% of those connections. "If all parts of the brain talk to all parts of the brain," says neuroscientist Ralph-Axel Müller, "all you get is noise." The theory is that this overcommunication creates a kind of neural cacophony that can make new sensory inputs like the voice of a stranger painful to absorb, rendering basic social functions nearly impossible. A recent study at Massachusetts General Hospital supports this theory for one of the hallmark traits of autism: eye contact avoidance. The researchers found that the subcortical system—responsible for recognizing others' faces and emotions—is overactive in individuals with autism, which may explain why they feel discomfort or even "burning" when looking others in the eye.

While there is no known cure for autism, scientists at SUNY Downstate Medical Center and beyond are starting to isolate the neural pro-

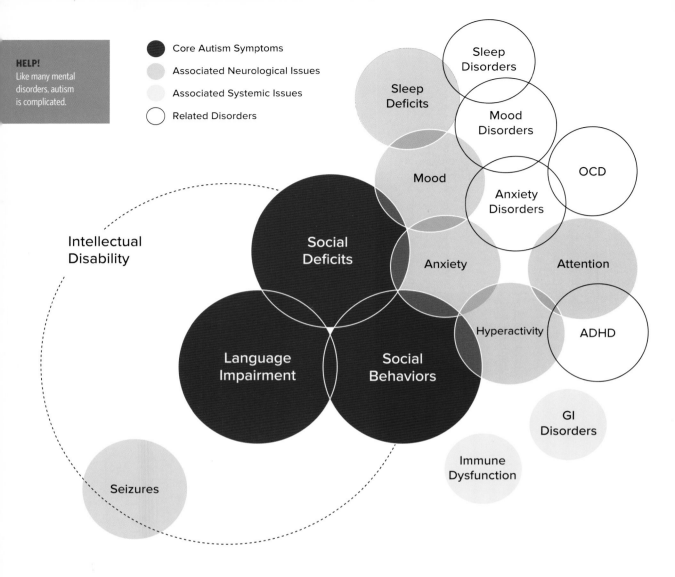

Core Autism Symptoms

Associated Neurological Issues

Associated Systemic Issues

Related Disorders

Sleep Disorders

Sleep Deficits

Mood Disorders

OCD

Mood

Anxiety Disorders

Social Deficits

Anxiety

Attention

Intellectual Disability

Hyperactivity

ADHD

Language Impairment

Social Behaviors

GI Disorders

Immune Dysfunction

Seizures

cesses and proteins that regulate the pruning process, which could soon lead to treatments for the estimated 1 in 68 U.S. children currently diagnosed.

Rethinking Our Approach to Brain Disorders

There's no question sufferers of autism spectrum disorder and similar conditions can face great hurdles interacting with "normal" society. But are we sure it's those on the spectrum who need to be reformed? After all, society has generally come around to accept other conditions once considered diseases, like homosexuality and women's "hysteria" (which seems to have been patriarchal code for not being properly docile). Some autistic people are capable of incredible feats of memory and focus, and might well be considered superbeings if not for the social awkwardness that attends their powers. Instead of working to bend autistic people's abilities to

society's norms, some say, it's time for society to become more inclusive.

This is the position of **neurodiversity** advocates, who question so-called cures for mental disorders that risk fundamentally changing who the patient is. Julia Bascom, director of programs for the Autistic Self Advocacy Network, explained her stance in *The Daily Beast*: "The idea of a cure for autism doesn't make sense. Autism isn't a disease or an injury; it's a neurodevelopmental disability that shapes our brains differently." Autistic people aren't rejecting social norms like handshakes and eye contact because they don't understand them but because their overworked brains can't easily incorporate them. "Is flapping my hands or intensely and obsessively loving something...the psychological equivalent of diabetes?" wonders Bascom, "or is it a natural and beautiful part of human diversity?"

Is There Such a Thing as a "Criminal Mind"?

Yes, unfortunately some people do seem to be at least partly predisposed to criminal behavior. Criminologist Kevin Beaver of Florida State University looked at the genetic profiles and criminal records of 1,000 young men. "There are some men that are, biologically and genetically speaking, much more likely to be predisposed to be violent and aggressive than others," he declared in the *NOVA* special "Can Science Stop Crime?" About a third of men have a variation of a particular gene that may make it more difficult for the prefrontal cortex to calm down your amygdalae when they're overly excited. In mice, manipulating this "**warrior gene**" can lead to rodents that are more aggressive, fighting and clawing at each other until they have to be manually separated. But it isn't the whole story—after all, clearly not all of those men are violent. The growing brain is very plastic, and a happier childhood environment and other factors apparently help build stronger and more reliable impulse control circuitry in the majority of those with the warrior gene. Worth noting: Psychiatric disorders by themselves are not a reliable indicator of criminality; those with disorders are 10 times more likely to be *victims* than perpetrators, according to the U.S. Department of Health & Human Services.

What Does Stress Do to Your Brain?

Chronic **stress** and the elevated levels of the stress hormone cortisol that accompany it have recently been found to upend the normal balance of new-brain-cell creation, leading to more myelin-producing cells and fewer neurons—artificially "hardening" some connections with extra myelin, so that chronically stressed people can be in a constant state of fight-or-flight. Stress can also shrink the hippocampus, which is critical for forming new memories; young people exposed to chronic stress have more learning difficulties and anxiety disorders. Bottom line: Finding a way to escape a source of ongoing stress, when you can, is a good survival strategy—trying to boldly suffer through it may do permanent damage to your mind and body. If you can't escape it, consider meditation and other forms of stress relief to try ameliorating the worst of the effects.

PSYCHO ANALYSIS

What's really wrong with these fictional villains?

JOFFREY BARATHEON,
Game of Thrones
DIAGNOSIS: **Conduct disorder**

This juvenile condition is characterized by a callous disregard for others, along with a history of hostility, violent behavior, and acts of cruelty. Children and adolescents with the disorder regularly lie, bully, cheat, steal, and abuse animals—and they somehow find it all gratifying, such as in *A Storm of Swords* where Joffrey is revealed to have killed and cut open a pregnant cat to see the kittens inside. This is often a precursor to an adult diagnosis of antisocial personality disorder.

THE JOKER, *The Dark Knight*
DIAGNOSIS: **Antisocial personality disorder** and **intermittent explosive disorder**

Antisocial personality disorder encompasses those who might otherwise be known as **sociopaths** or **psychopaths**. It's characterized by a lack of empathy and remorse and consistent patterns of deceit, aggression, and manipulation. The Joker is also thought to have intermittent explosive disorder, which is characterized by a very hot temper and hostile outbursts that are out of proportion to the perceived offense.

ALEX FORREST, *Fatal Attraction*
DIAGNOSIS: **Borderline personality disorder**

Borderline personality disorder is characterized by erratic fluctuations in mood and behavior, explosive bursts of anger, and intense, tumultuous relationships. One Australian patient describes it to Vice.com as feeling "like a tangled slinky, forever bumping inelegantly down a flight of stairs." Bunny-boiling aside, Alex Forrest has become synonymous with the disorder.

NORMAN BATES, *Psycho*
DIAGNOSIS: **Dissociative identity disorder**

While the diagnosis didn't yet exist at the time of the production, the psychiatrist's assessment in the film seems to support it. He determines that Bates' mind had adopted a "mother half" alter ego to distance himself from his matricide, and after committing each subsequent murder while in the alter ego state he would come to "as if from a deep sleep."

CRUELLA DE VIL, *101 Dalmatians*
DIAGNOSIS: **Narcissistic personality disorder**

She's arrogant, has a sense of entitlement, is cruel to others, and preoccupied with fantasies of beauty? Classic.

Do Electromagnetic Fields Harm Your Brain?

Electromagnetic fields, such as those generated by your smartphones, aren't good for your head. The World Health Organization (WHO) classifies them as a Class Two carcinogen, the same category as engine exhaust and lead. They can weaken your blood-brain barrier and increase cell-attacking "free radicals"; they've been linked to benign and malignant brain tumors after 10 years of exposure, with younger children especially susceptible to their effects. So use your cellphone on speaker mode or with headphones, and think twice about buying that cute house under the high tension wires.

Lights Out: What Is Death?

Is there anything as fascinating to the human psyche as death? But for all the apparent simplicity of it—you're not dead, and then you are—there's complexity. We often use the term "brain-dead," i.e., the ceasing of all neuron activity, as our boundary. But there are other choices and definitions, including **clinical death** (when the heartbeat and breathing stop), **legal death** (when the law considers you dead, by whatever statute applies), **biological death** (when your cells start to die off), and **information-theoretic death** (when your information—your preferences, memories, etc.—are no longer, in theory, recoverable).

"Clinical death" is trouble, for sure, when your heart no longer pumps blood and breathing stops. But your clever brain usually has a small cache of oxygen in reserve, and remains alive during this stage. In fact, a study of patients being removed from life support at George Washington University Medical Center noted a distinct spike in brain activity after blood pressure ceased, lasting anywhere from half a minute to three minutes, with the brain registering gamma waves during the cycle. That's usually a sign of a conscious brain; in this case, one might speculate, a brain frantically trying to figure out how to get the oxygen flowing again.

What happens next to your vast system of neurons normally in a steady state of electrical charge differentials is fairly amazing. As *Scientific American* reports: "As blood flow slows and oxygen runs out, the cells can no longer maintain polarity and they fire, causing a cascade of activity that ripples through the brain. If these seizures were to occur in memory regions, they could explain the vivid recollections often reported by people who are resuscitated from near death." A portion of the brain called the **locus coeruleus**, which is linked to the amygdala (emotion) and the hypothalamus (memory), releases a stress hormone called **noradrenaline** in times of trauma, which could combine with the spike of neural activity to flood the brain with highly emotionally charged memories. Similarly, the "light at the end of the tunnel" that resuscitated people claim to see might simply be tunnel vision caused by the cessation of blood and oxygen to the eyes.

You might assume the brain will nobly power on while it can after the heart has given up, but recent research reveals a rather stunning phenomenon: Researchers studying rats in the final minutes before they died noticed that at the end, the rats' brains essentially bombarded their hearts with signals that hastened the heart's surrender. It's almost as if the brain, upon realizing death is imminent, takes it upon itself to shut the whole show down. Diving in deeper, Jimo Borjigin, a neuroscientist at the University of Michigan Medical School, discovered that if you can keep a rat's brain from communicating with the heart during the death throes, the heart will actually keep beating and the rats will remain alive three times as long. Without interference from a panic-stricken brain, the heart could, in the words of Céline Dion, go on. It isn't just academic: Keeping the brain out of it during, say, a cardiac arrest, could potentially boost a patient's chance of survival—a concept that seems to go against all commonly held medical beliefs. Maybe the brain isn't entirely trustworthy, especially in times of crisis.

TIME TO GO
Six people died, somewhere in the world, in the time it took you to read this caption. So maybe pick up the pace?

"Death must be so beautiful. To lie in the soft brown earth, with the grasses waving above one's head, and listen to silence. To have no yesterday, and no tomorrow. To forget time, to forgive life, to be at peace."
—Oscar Wilde

SCREEN TEST

Did *A Clockwork Orange* get "aversion therapy" right?

We know we can associate behaviors with neutral stimuli; this is how Pavlov's famous dogs were conditioned to drool at the sound of a bell, after it was played a number of times whenever they were brought drool-inducing food. The aversion therapy portrayed in *A Clockwork Orange* takes this down a dark path, where fictional authorities purposefully associate undesired behavior (in this case, going on wilding rampages) with physical discomfort (here, a nausea-inducing drug). The idea is that if thinking about the behavior makes you sick, you'll avoid the behavior. So yes, the movie gets aversion therapy right, including its unintentional consequences—In the movie, Beethoven's music, played during training, induced the same nausea later; similarly Pavlov's dogs also drooled on seeing lab coats and hearing researchers' footsteps. But aversion therapy isn't considered effective and isn't widely used; in one study that tested aversion therapy by using electrical shock on alcoholics, those receiving the therapy performed worse than the control group 15 months later.

Life After Death

Further blurring the line between life and death, people can sometimes be revived long after their heart stops beating. "Biological death" occurs when your cells actually start to erode and die off, but that irreversible process can be slowed effectively to a halt by hypothermia.

Anna Bagenholm was skiing in a relatively remote area in Norway with friends in 1999 when she slipped and fell headfirst into a frozen river. She managed to find an air pocket beneath the ice and remained conscious for 40 minutes, but by the time rescuers got to her, she had been submerged for 80 minutes in the freezing water; she had no breath, no pulse, and a body temperature under 57° F—the lowest on record. But ER doctors managed to revive her, albeit with significant nerve damage, thanks to hypothermia. "Her body had time to cool down completely before the heart stopped," explained ER chief Dr. Mads Gilbert. "Her brain was so cold when the heart stopped that the brain cells needed very little oxygen, so the brain could survive for quite a prolonged time." The ability to preserve brain cells and connections through extreme cold is the basis for the science behind **cryonics**, the careful and hopeful preservation of either whole human bodies or just the noggin for "re-animation" by whomever colonizes Earth after us, à la Fry from *Futurama*. The jury is still out, but there have been advances: In 2016 a rabbit brain, with a complete **connectome** (all of its neurons and synapses) was successfully preserved for the first time ever, by a team led by MIT graduate Robert McIntyre.

If our bioscience skills continue to improve, it seems likely we'll reach a point where a preserved brain can be reimplanted into another body, Frankenstein-style. And that raises all kinds of questions about identity. How much of who we are is contained within the electrochemical activity and protein folds of your brain? Would preserving that structure really preserve you— would your consciousness, the intangible elements that make you *you*, survive the operation? Or would you be somebody else? One thing is for sure: We've only begun to tap the complexity that is you, as the next chapter, on intelligence, personality, and creativity, makes clear.

Is Walt Disney's head locked away in a cryogenic tube?
Nope. He was cremated and buried in Forest Lawn Memorial Park, in California. He showed interest in cryonics once, which seems to be where this curiously enduring myth came from. Sorry, people: There will be no Dumbo 2 from the master himself.

MYTHS FOR BUSTING

CHAPTER 6

LANGUAGE, INTELLIGENCE, CREATIVITY, AND THE MIRACLE OF GENIUS

Not to give you a big head, but you are one of the most intelligent creatures this planet has ever produced, in four thousand million years of trying.

Unless dolphins or dung beetles or Venus fly traps are philosophizing at some undiscovered fundamental level, it seems likely there's something about human intelligence that is not just better, but an order of magnitude more complex than that of any other life form on Earth. A parrot can repeat, but not compose; an elephant can be taught to paint, but not to *want* to paint. Intelligence is notoriously hard to define, but in our experience, we're the only creatures making the attempt.

Take that, dolphins!

MIND BENDERS

28 Age—in months—of the youngest member ever to join genius club Mensa

100,000 *Approximate age of human language*

1,000,000+ *Words in the English language*

It's estimated the average person knows 5,000 or so words at the start of grade school, and as many as 75,000 on leaving college.

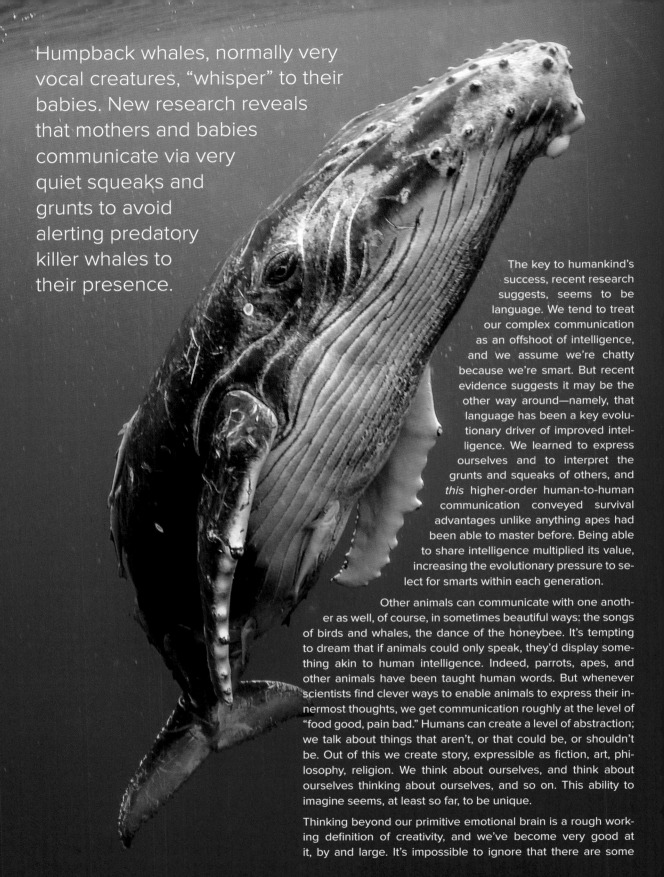

Humpback whales, normally very vocal creatures, "whisper" to their babies. New research reveals that mothers and babies communicate via very quiet squeaks and grunts to avoid alerting predatory killer whales to their presence.

The key to humankind's success, recent research suggests, seems to be language. We tend to treat our complex communication as an offshoot of intelligence, and we assume we're chatty because we're smart. But recent evidence suggests it may be the other way around—namely, that language has been a key evolutionary driver of improved intelligence. We learned to express ourselves and to interpret the grunts and squeaks of others, and *this* higher-order human-to-human communication conveyed survival advantages unlike anything apes had been able to master before. Being able to share intelligence multiplied its value, increasing the evolutionary pressure to select for smarts within each generation.

Other animals can communicate with one another as well, of course, in sometimes beautiful ways: the songs of birds and whales, the dance of the honeybee. It's tempting to dream that if animals could only speak, they'd display something akin to human intelligence. Indeed, parrots, apes, and other animals have been taught human words. But whenever scientists find clever ways to enable animals to express their innermost thoughts, we get communication roughly at the level of "food good, pain bad." Humans can create a level of abstraction; we talk about things that aren't, or that could be, or shouldn't be. Out of this we create story, expressible as fiction, art, philosophy, religion. We think about ourselves, and think about ourselves thinking about ourselves, and so on. This ability to imagine seems, at least so far, to be unique.

Thinking beyond our primitive emotional brain is a rough working definition of creativity, and we've become very good at it, by and large. It's impossible to ignore that there are some

HOWARD GARDNER'S
MULTIPLE INTELLIGENCES

In his seminal 1983 work *Frames of the Mind: The Theory of Multiple Intelligences,* Howard Gardner proposed an intelligence model with eight criteria, six more than the generally-accepted standard—the IQ test. While this theory has been debated within the scientific community, his work has impacted intelligence theory to this day.

Multiple Intelligences

LINGUISTIC
Spoken and written words; reading, listening, speaking, and writing

LOGICAL-MATHEMATICAL
Reasoning and problem-solving; numbers

MUSICAL
Songs, patterns, rhythms, instruments, and musical expression

BODY-KINESTHETIC
Interacting with one's environment; concrete, physical experiences

SPATIAL
Organizing ideas visually and dimensionally; thinking in images and pictures; "seeing" things in one's mind

INTERPERSONAL
Interacting with others; working collaboratively and cooperatively

INTRAPERSONAL
Feelings, values, and attitudes; empathizing with others

NATURALISTIC
Classifications, categories, and hierarchies; ability to pick up on subtle differences

among us who simply excel above the rest at creative and intellectual endeavors of all kinds. We call these people geniuses. "Genius" can be a slightly fishy term, and it can take many forms—academic, artistic, or even athletic. We even recognize evil geniuses and criminal masterminds. And there's a fascinating historical correlation between genius and madness that may offer clues to a deeper understanding of both.

The Origins of Language

Our earliest records of human language date from only 5,000 years ago or so, but that says more about our record keeping than our innate abilities. The human vocal tract, the differentiating hardware for our unique ability to speak, started developing around 100,000 years ago, with our mouths getting smaller, our tongues more flexible, and our throats longer. By 50,000 years ago or so, we seem to have evolved all the apparatus we need—including the tight breath control that makes long sentences possible—to make the complex set of sounds that set human speech apart. Evolution has built this incredible wind instrument not without some risk—your vocal cords sit choking-hazard-deep within your larynx (visible as the Adam's apple in most men and some women). But the small evolutionary risk is dwarfed by the advantages of being able to express things like, "Bob, cock your rifle—charging hippo on your left."

distinct brain regions, usually within the left hemisphere, that seemed mission-critical for language. Dr. Broca found that damage to part of the left frontal cortex (henceforth known as **Broca's Area**) could prevent a person from producing speech or forming words. Similarly, damage to **Wernicke's Area**, a spot in the posterior part of the temporal lobe closer to the rear of the brain, could make it hard to comprehend speech, and sometimes cause the afflicted to produce word salad: real words, but said in a scrambled order that made no sense. Together with the earlier research of French neurologist Marc Dax (who'd looked into aphasia, or the loss of speech capability), for the next hundred years, these results shaped our understanding about where language lives—including the widely held assumption that language is a left-brain function.

More recently we've come to understand that language, like complex thought, is distributed more broadly—in this case, all across the cerebral cortex. The results of a UC Berkeley study released in April of 2016, in particular, upends the notion of left-side dominance when it comes to processing words. The study exposed subjects to the autobiographical stories of radio program *The Moth*, while using fMRI imaging to trace the blood flow throughout subjects' brains. Researchers were able to map where the concepts representing more than 10,000 individual words were stored, defini-

A 2012 study found that dogs understand human communication **about as well as 6-month-old babies.**

It's easy to see how important language is to our species' success: Our capability for abstract thought allows complex cooperation, specificity of warnings and opportunities, inspiration through storytelling, and a million other advantages. But we've only been able to speak like this for about 50,000 years—a tiny blip on the evolutionary clock. How did humankind move from grunts to words to GIFs so quickly?

Early brain researchers looking for the seat of language in the head began their investigations as they so often do: by looking at brain damage. In the 1860s and 1870s, doctors Paul Broca and Carl Wernicke identified two

tively showing that the entire cerebral cortex participates.

Talk about voices in your head!

One particular form of language processing may occur far deeper in the brain than previously thought. You know it quite well—in fact, it's what you're doing right now.

The thalamus and brain stem help our visual cortex process what we're seeing, focusing on complex objects like faces and, as a recent large-scale study of illiterate Indian women suggests, letters. These ancient structures in our brains actually reprogram themselves to

Just how well do you understand how your brain processes language?
These four tests are designed to show just how far off your instincts may be.

FIND THE LIMITS OF YOUR LANGUAGE PROCESSING: FOUR TESTS

1

How You Actually Speed-Read

Reading might seem straightforward enough, but researchers found that we make a clever shortcut, paying attention only to the first and last letters and intuiting the middle based on context—which is why this might look like gobbledygook to a computer program, but is perfectly readable to you.

▼

It wsa mcuh plasentaer at hmoe wehn one wsna't awlays lraegr and smllaer, and bineg odreerd aubot by mcie and rabibts. I amlsot wsih I hdna't gnoe dwon the rabibt-hloe.

From *Alice's Adventures in Wonderland* by Lewis Carroll

Light

Light house

Light housecleaning

Set

Set the table

Set the table on fire

Fly

Fly up

Fly up your nose

2

How Context Determines Reality ▲

One word can have multiple meanings, making the simple act of reading a constant reassessment based on the words around it. Consider how you feel about the highlighted word as you move down each short list.

RED BLUE GREEN PURPLE BLACK ORANGE BROWN

3

How Your Brain Handles Simultaneous Tasks ▲

Despite popular belief, we don't multitask the way that we think we do. The brain cannot process two "conscious" activities—texting while walking or driving, say—simultaneously. What we're actually doing, according to MIT neuroscientist Earl Miller on NPR.org, is bouncing our focus back-and-forth between the two tasks. "Switching from task to task, you think you're actually paying attention to everything around you at the same time. But you're actually not." What your brain can do, however, is pick up on supporting stimuli while you're busy doing one task. Consider reading: As you read, your mind looks for contextual clues beyond the words. To see your overactive multitasker for yourself, just try reading the color of each word aloud, in order, without being influenced by the meaning of the word.

▲ Which one are you?

4

How Language Follows Biology ▶

Many linguists believe language began as onomatopoeia: words that reflect a sound, like "bang" and "oomph." Can you correctly infer the meanings of each of these words you've probably never seen?

ANSWERS

1. Round 2. Bright 3. Small 4. Down

1. YORUBA
Baluma means a **round** or a **sharp** object?

2. TAMIL
Olimikka means **dark** or **bright**?

3. GUJARATI
Jhiini means **big** or **small**?

4. DUTCH
Neerwaarts means **up** or **down**?

help us translate letters of the alphabet into language. And the more efficiently these structures communicate with our visual cortices, the more proficient we'll be at reading.

Is Animal Communication the Same Thing as Language?

Our animal friends may not have a mission to Mars in the works, but they clearly communicate. Listen to the dogs in any dog park, or the howling of wolves in the wild, or picture the dance by which bees convey the location of nearby flowers to a hungry hive. Is animal communication different from human language? If so, does it vary only by degrees, or is there a fundamental difference? With time and training, could a monkey compose a Shakespearean sonnet that another monkey could critique?

There've been a number of promising examples of individual animals that, with patient training, display something like intelligent communication. There was Koko, the gorilla who learned to understand about 2,000 English words and could communicate 1,000 words in sign language. (Sign language bypasses the hurdle caused by the fact that many animals' bodies aren't built for speech. Using gesture and movement lets us get at the more interesting question of whether they are capable of using language.) Chaser, a border collie, learned the names of more than 1,000 different toys and could retrieve them from another room based on the spoken word alone. Alex, a famous African grey parrot, *could* speak (with no lips, parrots reproduce human sounds like "P" and "B" with a unique flute-like organ called the syrinx). He displayed knowledge of some 100 words and could count to six. Clearly, these animals had something special going on.

But repetition is a far cry from conversation. Chaser's trainer worked with her five hours a day for five days a week for years, and the dog's intelligence at the end of her training has been described as equivalent to that of a two-and-a-half-year-old toddler. Impressive, by dog standards...but still. Stan Kuczaj, a psychologist from the University of Southern Mississippi who'd worked hard to develop communication with dolphins at Walt Disney World's Epcot, complained that the dolphins "were only really interested in communicating about needs that they had, like a tool they needed or a fish they wanted. We hoped they would also comment on other things going on in the aquarium

but they didn't." On further examination, the speech of Kanzi, a sign-language-using bonobo was revealed to be 96% merely functional (i.e., requests for food). There's been consistent criticism that human caretakers/interpreters of these animals are either leading them (consciously or unconsciously) to predetermined responses, or reading too much into what they seem to be saying.

In short, animals communicate—but they don't seem to have the level of abstraction that allows creativity. Their communication can be deceptively complex: Killer whale calves learn calls and specific sounds from their mothers, sounds that grow and mature as they do, and a 2014 study published by the Royal Society found other animals with complex vocal models, including free-tailed bats and Bengalese finches. To the extent we can understand it, their communication seems to be very immediate and physical, operating largely at the instinctual level of needs and fears, which is the older, reactionary "lizard brain" at work.

TO THE LETTER
Longest and shortest alphabets in the world? Cambodian: 74 characters. Rotoka (Papua New Guinea): 12 characters.

Using language involves learning a set of rules that can be applied to express complex ideas about never-before-encountered situations. But animals don't seem to be doing any of that; they're essentially just pointing, using sounds instead of paws or beaks.

Human language is infinitely more nuanced and abstract—we take words and ideas, strap them into rules of grammar and syntax so we're all on the same page, and spin out infinite combinations. Our ability to consider things in an abstract way lets us apply meaning to words or combinations of words apart from the immediacy of the moment. Humans can talk about things that aren't and things that should be. We can imagine the future and reflect on alternative pasts.

Animals can grasp concepts like "go" and "get" and "eat," but they don't seem to have much interest in anything more complex than that. Our conversations with dolphins, great apes, and the very smartest dogs are destined

COLLOQUIAL KLINGON

Fictional worlds need fictional languages, and for veracity a lot of attention is paid to rules and structures that make them feel consistent and "real." Some sample phrases in a few of mankind's recent inventions:

NA'VI (from *Avatar*)
Na'vi: "Taronyut yom smarìl."
English: "The prey eats the hunter; everything goes wrong that can."

DOTHRAKI (from *Game of Thrones*)
Dothraki: "Hash jin zhori ray efesash hrazefoon fini nem dranesh she ram ma fini nem azhish vigoverat?"
English: "Are these hearts from pasture-raised, sustainably-farmed horses?"

SIMLISH (from *The Sims* videogame series)
Simlish: "Nicloske Ga Gloope."
English: "I would like a taxi to come pick me up."

QUENYA ELVISH (from *The Lord of the Rings* trilogy)
Elvish: "Elen síla lúmenn' omentielvo."
English: "A star shines on the hour of our meeting."

KLINGON (from *Star Trek*)
Klingon: "mo'Dajvo' pa'wIjDaq je narghpu' He'So'bogh Sajllj."
English: "Your stinking pet has escaped from its cage and appeared in my quarters."

LATKA (played by Andy Kaufman in *Taxi*)
Latka-ese: "Ha-Dee-Fee-Bee"
English: "Revolution"
Latka-ese: "Ibi-da"
Ibi-da: "That is right"

to be shallow: not because the animals can't communicate, but because they have nothing interesting to say.

We can't seem to stop trying, though.

The Mysterious Origin of Human Language

Why are humans the only species to develop this complex language capability? What was it that elevated early hominids—our human ancestors—from animal-level screeching and hooting to being able to communicate abstract concepts like frenemies, space travel, and the meaning of "covfefe?"

One theory: A genetic mutation in hominid brains, after our break with other great apes around six million years ago, sped up the process of human brain evolution. The hopeful search for a specific "language gene" has been going on at least since the 1990s, when a mutation of the adorably named forkhead box protein P2 (a.k.a. FOXP2) seemed like a tantalizing candidate. Genetic research showed that out of 37 members of a Pakistani extended family, nearly half suffered from the same severe speech disorder. McGill University linguist Myrna Gopnik contended that a mutation of the "FOXP2" gene was the cause of their particular difficulty (dealing with tenses and plurals). Was this the language gene? Her theory was celebrated in some circles, but dissenters like neuroscientist and language expert Faraneh Vargha-Khadem condemned Gopnik's theory as "misleading" and "inaccurate," noting the affected family members also had orofacial issues like extreme difficulty closing one eye or clicking their tongues.

Other scholars surmise that our language really began as body language and primitive sign language, noting that we still gesture when we speak. The vocal tracts of early hominids couldn't form the same sorts of sounds modern humans can and, according to this theory, complex concepts were conveyed through signs. As Ray Jackendoff suggested in a report titled "How Did Language Begin?" for the Linguistic Society of America, the question is really what aspects of language are unique to humans and what aspects simply draw on other human abilities not shared with our other primate relatives. "Some researchers claim that everything in language is built out of other human abilities," notes Janckendoff. "The ability for vocal imitation, the ability to memorize vast amounts of information (both needed for learning words), the desire to communicate, the understanding of others' intentions and beliefs, and the ability to cooperate." In other words, according to this theory, language isn't itself unique—it's just a combination of other uniquely human capabilities.

A more recent theory suggests that the reason we developed language while other apes did not is because our brains got fat. Specifically, according to a recent study conducted by the Max Planck Institute for Evolutionary Anthropology, fat molecules called **lipids**, important for enabling neural activity in the brain's **neocortex**, evolved much more quickly and with more variety in human brains than in those of chimpanzees, though we split from our common ancestor at around the same time. The proliferation of these lipids may have been as important to the development of our complex brain function as anything on the genetic level. It's not bad to have a fat head.

Ultimately, the question of why humans uniquely developed language, shrouded as it is in our prehistoric past, may never be known for sure. In 1866, the French Academy grew so tired of all the baseless speculation that they banned all papers on the origin of language.

Does Speaking Multiple Languages Make You Smarter?

One common proxy for someone's intelligence is the number of languages they speak; learning a new language is such a daunting challenge that polyglots and hyperpolyglots (who speak six languages or more—a skill not for the timid) are often assumed to be intellectual overachievers. And it turns out their bragging rights are at least somewhat justified: Bilingual people may, in fact, develop bigger brains. A 2012 study of a group of cadets in the Swedish Armed Forces' Interpreter Academy found that cadets learning new languages—Arabic,

Russian, or Dari—showed actual physical growth in the size of language-related parts of their brains like the hippocampus, which, in this case, helps you anticipate how a sentence is going to unfold. It wasn't just *studying* that did it—medical students studying subjects other than language showed no such growth.

Speaking more than one language can convey other benefits as well, including—incredibly—postponing the effects of degenerative brain diseases like Alzheimer's. An Italian study, for example, found that bilingual Alzheimer's patients had increased neural connectivity in frontal brain regions, which helped sustain their performance on memory tasks.

Interestingly, the language you speak can affect the way you perceive the world. In 2014, the journal *Psychological Science* published a study that revealed differences in the way speakers of different languages understood the same external stimuli. Asked to describe a video clip of a woman walking through a parking lot toward a car, English speakers focused on the simple action ("The woman is walking") while German speakers tended to supply imagined intention ("The woman is walking toward her car.") Even more intriguingly, bilingual German/English speakers focused on the action (i.e., the English way) when speaking English, but on the intention (i.e., the German way) when speaking German. It's intriguing evidence that our language helps organize our thoughts.

Can Emerging Tech Remove the "Language Barrier" Once and for All?

Almost certainly. Science fiction has long cheated on this issue by casually making alien species speak broken English, or inventing technology that instantly gives any user the power to communicate directly with whomever they encounter on their adventures (like *Star Trek*'s Universal Translator, or the ear-implanted "Babel fish" in *The Hitchhiker's Guide to the*

"Great wits are sure to madness near allied, and thin partitions do their bounds divide."

—John Dryden, English poet & playwright

"Talent hits a target no one else can hit. Genius hits a target no one else can see."

–Arthur Schopenhauer, German philosopher

Galaxy). As our world becomes more and more globally minded, and with more businesses featuring remote offices and untethered full-time staff all over the globe, efforts have redoubled to use tech to leap language barriers in a single bound.

Already you can type any phrase into Google and have it instantly translated into more than 100 languages; Microsoft, in December of 2016, launched a Universal Translator app that connects up to 100 people via their smartphones and translates their spoken words into one another's languages. A company called Waverly Labs is developing a small, hearing aid-sized device they've dubbed "The Pilot System," which instantly translates English, French, Spanish, and Italian in real time (eventually they hope to include Slavic, Semitic, Hindi, and East Asian languages as well). This would be, essentially, the "Babel fish," but sleeker, less invasive, and available in fashionable colors.

The only price for tech-driven universal communication is that you'll deny yourself the improved mental functioning you'd gain from learning additional languages. (Similarly, when deaf children of deaf parents get cochlear implants at an early age, they skip the most fertile period for absorbing sign language, and will effectively not speak their parents' language.)

Brain-boosting tips for learning a language:

1. Soundtrack your sleep. A Swiss study found that German speakers who studied Dutch before bedtime, then played language tapes during their non-REM sleep, performed better on vocab tests.

2. Talk with your hands. German scientists teaching a made-up language found that gestures improved students' retention of the language. Surprisingly, even arbitrary gestures worked for abstract words and adverbs.

3. Get reading. It's far easier for the brain to learn to read a new language than it is to speak one.

The Rise of Intelligence

With language as a driving force, and civilization imposing some helpful constraints on our natural simian social patterns, *Homo sapiens* started dominating, developing agriculture, establishing permanent dwellings, and wiping out those pesky Neanderthals. The specialization that civilization allowed began to give humans the opportunity to explore myriad things. Writing, art, politics, history, science, and everything else we know as being distinctly human was underway.

But first: What *is* smart? What *is* intelligence?

If intelligence is defined as the capacity to learn and the ability to apply knowledge, then the idea of comparing relative intelligence—one person's intelligence to the next's—is, accordingly, a complex concept to grasp. What you apply *your* intellect to is different than what I apply *mine* to; therefore, it's difficult to define who's really smarter.

For various reasons, but mostly just to judge and label one another, we've long tried to quantify intelligence through testing. Unfortunately, these measures of intelligence are still too simplistic: An intelligence test can gauge how well you perform in different tasks, including mathematical ability, analytical thinking, spatial reasoning, and logic, but whether any or all of those add up to generalized "intelligence" outside the narrow test setting is still anybody's guess: You may, or may not, be smarter than your little brother. And the most famous of these, of course, is the **IQ**, or **Intelligence Quotient**, test.

In the early 1900s, Alfred Binet, a self-taught psychologist, began developing a series of questions that would eventually evolve into what we know as the Stanford-Binet IQ test. The basic idea was that adaptability and judgment were better determinants of intelligence than "book smarts." Along with collaborator Theodore Simon, Binet wrote in 1916, "A person may be a moron or an imbecile if he is

WHAT THE HELL HAPPENED TO **ALBERT EINSTEIN'S BRAIN?**

Albert Einstein's dying words are lost to history; he mumbled something in German, a language his attending nurse couldn't understand. But the story only got crazier from there for the 20th century's wild-haired, beloved genius. He was cremated the same day he died, but the doctor who performed his autopsy, Thomas Harvey, thoughtfully and without permission from the family, removed the great man's brain before the cremation, for scientific study. Then he held onto the brain for the next 40 years. It started reasonably enough, with Dr. Harvey responsibly measuring and photographing the brain from all angles, but it ended badly; by the time he was in his 80s, he had lost his job and his marriage and was living in Kansas, using a cheese board to carve off slices of the greatest brain of the 20th century for those who requested them. At last, he gave what was left to the Princeton Hospital pathology department, from which he'd been removed 40 years earlier. The brain slices, on slides, now reside in the Mütter Museum in Philadelphia.

So...was Einstein's noodle any different than yours and mine? Yes and no. The brain wasn't particularly big—in fact, according to Harvey's initial weighing, it was slightly smaller than average. But later research would uncover a few intriguing differences. One study showed his brain to have a slightly higher than usual concentration of glial cells, which feed neurons and form the insulating myelin that promotes fast neuronal communication. Another showed a tighter packing of neurons in his cortex, possibly an indicator of faster processing, and there were a few surface anomalies, including asymmetric parietal lobes, four ridges on his prefrontal cortex (where working memory lives—most of us have only three) and a "Sign of Omega" associated with musicians (Einstein played the violin) atop his noggin. In other words, nothing obviously superhuman—just a particular, unique pattern that added up to one of the greatest minds that ever lived.

> **"Being smarter gives you a tailwind throughout life.** People who are more intelligent earn more, live longer, get divorced less, are less likely to get addicted to alcohol and tobacco, and their children live longer."

–Stephen Pinker, Cognitive scientist at Harvard Universty

lacking in judgment; but with good judgment he can never be either. Indeed the rest of the intellectual faculties seem of little importance in comparison with judgment."

Binet's tests, which were conducted with children, functioned under the assumption that intelligence developed as a function of age. He didn't develop the actual concept of an IQ number, or intelligence quotient—that came later, from a German psychologist named William Stern. Stern's theory was that physical age and mental age didn't always go hand in hand, and the IQ number was derived by taking the subject's physical age and dividing it by the "mental" age determined by his questioning, then multiplying by 100. An IQ of 150 would therefore declare you 1.5 times as intelligent as your age would predict, would put you in the top two percent of humans, and would easily qualify you to join **Mensa**, the famous high IQ society.

Is it valuable? A high IQ score has long been cited as a good predictor of long life, health, education, and income level. Some studies indicate that intelligence may be as much as 85% heritable, and your IQ may additionally be strongly affected by the environment and culture in which you were raised. Poverty negatively impacts IQ scores, for example: As NYU sociology professor Patrick Sharkey wrote in *The American Prospect*, "Living in poor neighborhoods over two consecutive generations reduces children's cognitive skills by roughly *eight* or *nine* [IQ] points ... roughly equivalent to missing *two to four years of schooling*." Also interesting: Breast-fed babies gain a few IQ points over formula-fed babies.

As Illinois State University psychologist W. Joel Schneider pointed out in *Scientific American*: "High IQ is nice to have and there is abundant evidence that it is substantially correlated with creative productivity. On the other hand, many people with high IQ fail to create much of anything and many people with moderate intellect achieve lasting greatness." Your results may vary.

Hoping for a more statistical basis for comparing intelligence, neurology researchers at UCLA in 2009 used modern imaging techniques to measure brain processing speed in twins. They used diffusion tensor imaging to paint a picture of the brain's wiring as subjects took a standard IQ test, and watched how quickly the brain responded to challenges. They discovered that brain processing speed correlated with the structural integrity of the white matter (the axons that carry signals from one part of the brain to another), with stronger white matter correlating with faster processing and higher intelligence. Since white matter formation is genetically determined, this could explain why intelligence seems largely heritable. One encouraging finding from the UCLA study is that brain improvements and developments can continue well into adulthood—learning a new language or taking up an instrument can improve your brainpower, even later in life.

More recently, a UK study found that IQ and creativity across cultures are positively correlated with cognitive "variability" and "adaptability;" that is to say, how frequently and adeptly different brain regions connect and reconnect to each other.

Beyond the IQ, you might also want to assess your **Emotional Intelligence**, an idea apparently coined in 1964 but reinvigorated in 1995 by author and psychologist Daniel Goleman. His controversial book, *Emotional Intelligence: Why It Can Matter More Than IQ*, aims to quantify the ability to recognize your own and others' emotional states along the axes of self-awareness, self-regulation, social skill, empathy, and motivation. EI and EQ—**Emotional Quotient**—are somewhat debated concepts in the research community, with some seeing value and others arguing that EI is merely a subset of intelligence itself. Professor Edwin A. Locke of the University of Maryland, for example, argued in 2005 that EI is simply a skill in which the ability to grasp abstractions is applied to emotions. The theory is that people who possess a high EQ are better leaders because they're more dependable, engaging, and empathetic; on the darker side, with their above-average ability to understand emotional cues, they may be more adept at manipulating others, or at getting people to act against their own interests by targeting emotional responses.

What Is Creativity?

"Creativity is piercing the mundane to find the marvelous," said journalist Bill Moyers. It's seeing something that nobody else sees and framing it in such a way that people can see it, whether it's creating a cucumber shark for Pinterest or reimagining the experience of driving a car. Though we prize creativity in culture, in business, and in relationships, it can be tricky to define— and while we're all capable of bursts of creative insight, some seem to excel at it. Or maybe creative types like John Lennon and Wes Anderson are simply those people brave

enough to allow themselves not just to think, but also to live, outside the box. "Creativity is allowing yourself to make mistakes," says *Dilbert* cartoonist Scott Adams. "Art is knowing which ones to keep."

Recent research suggests that a lot of what we know as creativity comes down to simply *not* being able to focus. Research conducted by Northwestern University in 2015 found evi-

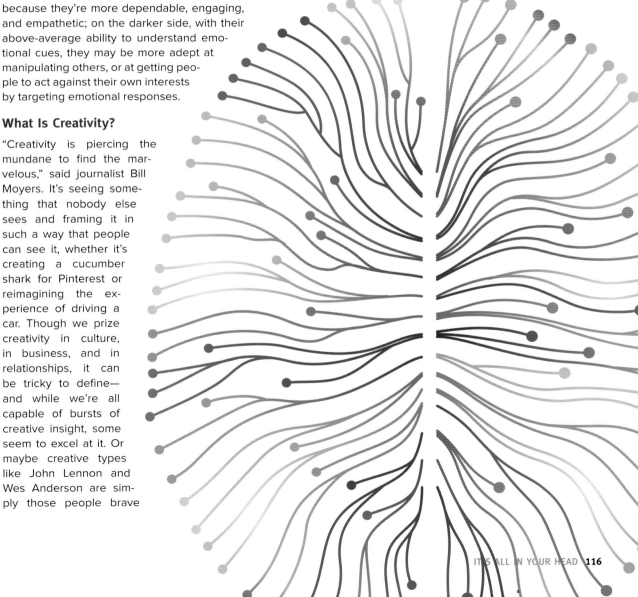

dence that the minds of creative people have a reduced ability to filter out irrelevant information. This leaky screening happens early in the brain's development, and is involuntary, but it enables certain people to entertain ideas outside of their immediate sphere of focus and incorporate elements that the general run of people dismiss.

Creativity is a sort of flexibility of the mind. It can be tied to neuroplasticity, the ability of the brain to change its structure over time, and to a related concept called "cognitive flexibility," used by psychologists to describe your ability to shift from one train of thought to another. Travel, for example, with its constantly shifting situations and contexts, has been shown in multiple studies to correlate well with improving creativity, particularly when the traveler is immersed in foreign cultures. "Foreign experiences increase both cognitive flexibility and depth and integrativeness of thought, the ability to make deep connections between dispa-

rate forms," says Adam Galinsky, a professor at Columbia Business School.

Creativity has often been tied to anxiety as well—ask any stand-up comic. The mind that can invent reasons to panic in the absence of a definite threat is a mind teeming with imagination, and some of our greatest creative minds have been among those the most ill at ease: David Letterman, Tina Fey, and many others. "In a sense, worry is the mother of invention," says Dr. Adam Perkins of King's College, London, in *Higher Perspective*. "Many of our greatest breakthroughs through the years were a result of worry. Nuclear power? Worry over energy. Advanced weapons? Worry of invasion. Medical breakthroughs? Worry over illness and death."

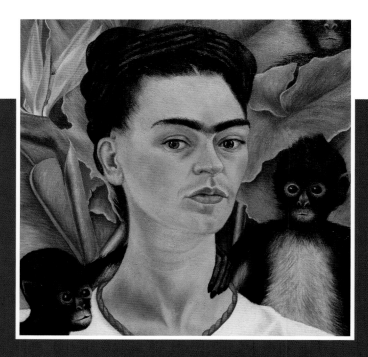

THE BIZARRE HABITS OF GENIUSES

You don't have to be crazy to be a genius...but it helps to have quirks.

Beethoven: Used to pour water all over himself while composing.

Prince: Savored unorthodox food pairings, like spaghetti and orange juice.

Nikola Tesla: Did 100 toe curls with each foot before bed to "boost his brainpower."

Lord Byron: Kept a bear in his dorm room and tried to get it a fellowship at Cambridge.

Edgar Allan Poe: Married his 13-year-old cousin (he was 27); wrote his works on long, thin scrolls.

Charles Dickens: Compulsively combed his hair hundreds of times a day.

Jean-Michel Basquiat: Painted in oversized Armani suits, which he would then wear out on the town still covered in paint.

Lucille Ball: Hoarded pencils.

Albert Einstein: Once plucked a grasshopper off the ground and ate it.

Frida Kahlo: Kept a jar with a formaldehyde-preserved fetus on her bedroom bookshelf.

Pythagoras: Reportedly starved himself, sometimes for 40 days at a time.

Tycho Brahe: Kept a dwarf under his dinner table.

Others who do not filter stimuli particularly well are those with psychotic disorders like schizophrenia and those on "magic mushrooms" and other drugs containing **psilocybin**. Creativity is a gift with some serious strings attached: One minute you're innocently painting starry nights, and then next you're delivering your ear to a brothel.

What Is Genius?

For as often as we throw the word around— "Whoever thought of chili cheese dogs was a genius!"—there's no easy definition for *genius*. Not even Mensa, the international organization specifically devoted to the highest IQs in the world, can do it: Most members don't refer to themselves that way, and credit their membership only to doing well on specific tests. In an essay titled "The Probability of Genius," psychology professor and author David Lubinski wrote that genius "is such a rare phenomenon, some have questioned whether it is meaningful to attempt to study it scientifically." Plus, as he goes on to contend, genius comes in so many flavors it would be difficult to know where to start. There are mathematical geniuses, artistic geniuses, criminal geniuses.

Neuroscientist Claudio Del Percio of Sapienza University in Rome conducted a study in 2009 to test whether "athletic genius" was a viable and definable entity. Del Percio and his colleagues measured the brainwaves of athletes and non-athletes at rest and during activity, and found that athletes' brains were honed and tuned like sports cars. At rest, they emitted stronger alpha waves, signaling a readiness to leap into action immediately, but during moments of activity their brains didn't work as hard as non-athletes. In other words, their brains were quieter and more efficient, and didn't have to devote as much brainpower to motor functions—efficiency perhaps related to their more highly attuned muscle memory. Genius!

One common assumption about genius is that it tends to manifest itself at a young age. Pablo Picasso could focus on nothing but drawing and painting as a child (according to popular history his first words were Spanish for "Pencil! Pencil!"). Marie Curie taught herself to read French and Russian by age 4. Wolfgang Amadeus Mozart wrote his first symphony when he was 8 years old and at age 6, future Manhattan Project physicist John Von Neumann could already multiply eight-digit numbers in his head. There are of course countless counterexamples of genius come late—Albert Einstein himself was still a patent clerk heading into age 30. But child prodigies make compelling stories.

There apparently is something to the notion that geniuses are born, not made. In a National Institute of Mental Health study, Dr. Judith Rapoport and team followed students from the 1980s until 2006, and found that their brains' development peaked at different times. The team expected to discover that earlier cortex thickening among 4- to 8-year olds meant a higher IQ; however, Dr. Rapoport was surprised to discover that the opposite was true—the younger students with higher IQs actually had thinner cortices. "The very, very brightest third had a later development of their cortex," noted Rapoport. "At the earlier ages they hadn't started really growing [i.e., thickening] to their peak amount yet." Simply put, as a group, kids whose cortices started thickening later eventually outperformed their peers intellectually.

What's your definition of "genius?"

The Curse of Genius

Geniuses come in all shapes and sizes, but although there's no scientific proof, history has taught us that genius and tortured go all too easily hand in hand. For more than a few of our most exceptional minds, their gifts seem to have come bundled with odd, eccentric, and even antisocial tendencies. Take these two examples of so-called tortured geniuses:

WILLIAM JAMES SIDIS

Born in 1898 to Ukrainian immigrants, Sidis came from intelligent stock: His father was a psychologist who dabbled in hypnosis, and his mother, among the first women to obtain a medical degree, was so intent on molding young William into a "genius" that she quit her job and devoted the family's life savings to that very task. William could speak at 6 months old and feed himself at 8 months; by the time he was 2, he was reading *The New York Times*; by four he could reportedly type in both English and French. "Emotional immaturity" was the only thing keeping him out of Harvard at age 9 (he would eventually get in at age 11), and that problem seemed to persist: First he quit teaching mathematics, finding it too difficult to be younger than his students; then he dropped out of Harvard Law School; eventually Sidis spent a lot of time in jail for organizing anti-draft and pro-communist rallies. After living in seclusion for some time, William eventually suffered a massive stroke and died at age 46.

SRINIVASA RAMANUJAN

Ramanujan, the subject of the 2015 film *The Man Who Knew Infinity*, was born in a small Tamil village in 1887 and excelled early on in a variety of subjects. By the time he reached university, however, he was so single-minded about mathematics he failed his other subjects, lost his government grant, dropped out of school, and applied for unemployment. A mathematic autodidact—one who is self-taught—Ramanujan worked off his own formulas and with his own methods, often breaking new ground. His seemingly preternatural ability with math brought him to Cambridge at age 26 to study with famed English mathematician G.H. Hardy. But Ramanujan contracted tuberculosis, was confined for a while in a sanatorium, and returned to India at the age of 31, where he died, leaving behind notebooks filled with strange and puzzling mathematical equations.

The "tortured genius" is a very compelling idea, and has remained a staple of pop culture across the span of human history. But are creative genius and mental illness truly linked? The science is not yet settled. Some experts, like clinical psychologist Kay Redfield Jamison, point to dozens of studies correlating creative genius with mental disorders, particularly with bipolar disorder. A burst of creativity and the mood improvement of a bipolar sufferer, for example, are marked by a similar shift in brain activity—indicating a connection, to some. Or as mental health law professor Elyn Saks puts it: "I think the creativity is just one part of something that is mostly bad."

But others are not convinced. "Creative people are not more likely to be diagnosed with mental illness, and mentally ill people are not more likely to be creative than normal people," says UNC professor Keith Sawyer in a scathing review of comparative studies. He and others believe the research is flawed by small sample sizes and self-reported measures of elusive "creativity." For them, whatever correlation we see is accounted for by the law of large numbers: Mental illness of one kind or another is depressingly common—nearly half of us will experience it at some time in our lives, according to a recent large-scale study of nearly 10,000 Americans—so no matter how rare a genius is, odds are that many of them will coincidentally also be suffering from mental illness.

Is It Possible to Raise a Genius Baby?

Remember all those Baby Mozart CDs, and the notion that babies would be born smarter if we just aimed classical music at their womb? Yeah...about that.

It was the 1993 work of psychologist Frances Rauscher and colleagues, whose college-age subjects showed improvements in spatial thinking and problem solving after listening to classical music, that led directly to prenatal belly headphones and the national craze for genius babies. Rauscher had never intended for her work to be applied to the prenatal world, but an overinterpretation of the results of Rauscher's tests quickly spiraled out of her control, to the point where the governor of Georgia issued a mandate that all expectant mothers were to be issued classical music CDs, and the state of Florida demanded that classical music be piped into daycare centers. Research published in the *Journal of Pediatrics* indicated playing Mozart to your babies might

WHAT KIND OF PARENT ARE YOU?

A University of Texas at Austin study of nearly 500 Chinese-American families evaluated how four parenting styles—"Tiger" (controlling and punitive, with high academic expectations), "Harsh" (controlling and punitive, with low academic expectations), "Supportive," and "Easygoing"—influenced adolescent development across eight years. Researchers found that "Tiger" parenting led to lower grade point averages and higher susceptibility to depression in teens than "Supportive" parenting. You may be familiar with this concept from Amy Chua's 2011 bestseller *Battle Hymn of the Tiger Mother*, but despite common generalizations, "Tiger" parenting isn't as common among Chinese-American families as is often assumed, and is actually declining in favor of "Supportive" parenting.

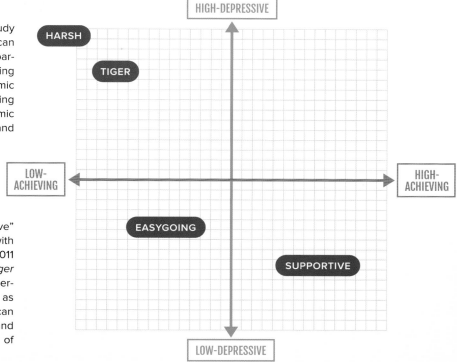

actually make language development slower. "I think parents are very desperate to give their own children every single enhancement that they can," Rauscher lamented.

Is there anything you can do to improve your unborn child's intelligence? Well, as we've seen, intelligence seems to be 85% inherited, so most of it's out of your control. If breastfeeding is an option, you could get a boost there: Children who are breast-fed for longer than six months have been shown to have IQs 3.8 points higher than children who are bottle-fed. What mom eats during the breastfeeding period is important as well, and you can boost baby's brain-building with salmon for omega-3 fatty acids, blueberries for antioxidants, and chicken or lean beef for iron. And *after* the baby's born, brain development is still in full swing—so be sure to provide lots of one-on-one time and baby talk—babies pay more

attention to that "goo-goo" talk than to background adult conversation, and it can help their language development. (Babies who get more baby talk know more words at age 2, according to WebMD.)

A more recent baby craze resulted from author Amy Chua's 2011 book *Battle Hymn of the Tiger Mother*. The book celebrated the discipline-heavy parenting techniques of Asian parents, and it not so subtly claimed that Western moms were too laid-back and therefore were raising underachievers. A Stanford University study revealed that "cool" moms, "tiger" moms, and those in between had almost equally effective influence, and that the difference in results mostly boiled down to how children viewed their parents through the prism of the culture they were raised in. (Asian children tend to look to parents for support throughout their lives, while European and American children are more likely to prize their independence.) Other studies, covered in the American Psychological Association's *Asian-American Journal of Psychology*, showed that tiger parenting wasn't as common among Asian parents as the book implied, and that where it was employed, it could sometimes lead to an increase in depressive symptoms and alienation from parents.

Intelligence, creativity, and genius are fluid concepts on a continuum that resists easy demarcations. There's no question some individuals have exceptional minds, but why and how this happens isn't very well understood yet, and trying to link success to genes or specifics of brain structure hasn't been very fruitful so far. But you are a quantum leap smarter than even your smartest border collie friends, and you can thank your generally expanding cerebral cortex— and the development of language in particular—for that.

Your ability, perhaps unique in nature, to have abstract thoughts and engage in self-reflection makes a much richer and imaginative inner world possible; it allows the cooperation and specialization that turned family into civilization, and this in turn led rapidly to all kinds of gains, not just in the practical realms of trade and technology, but in the squishier things we now hold to be very important, like art, culture, romance, and philosophy. We anoint some among us geniuses in all kinds of cortex-intensive realms, even though it's unclear what constitutes a genius or whether that distinction even makes sense.

In short, we really don't know what the upper boundaries are in terms of our highest orders of thinking; it may not be knowable. But an insatiable curiosity about how our minds work has been with us since the earliest days and has led to a curious human phenomenon: We regularly attempt to alter our minds, through various substances, activities, and experiences designed purposely to hijack our most basic risk and reward systems and to see what else we're capable of. One popular shorthand for these brain benders is sex, drugs, and rock & roll...and that's the subject of the next chapter.

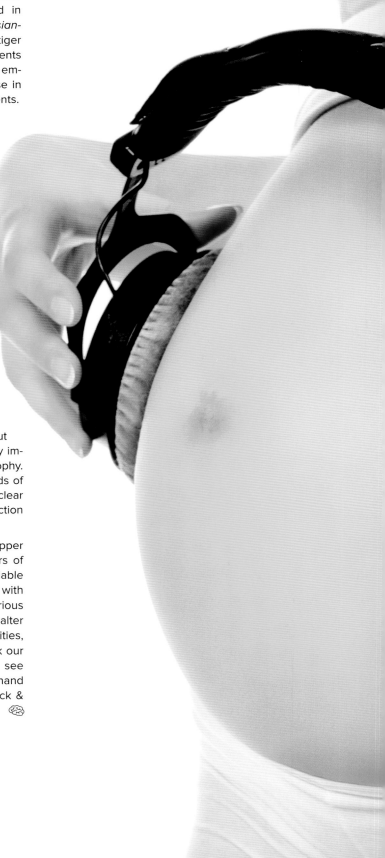

Studies have shown exposing babies to music is beneficial to their development, but prenatal exposure is thus far inconclusive.

CHAPTER 7

ALTERED STATES:
YOUR BRAIN ON SEX, DRUGS, AND ROCK & ROLL

Let's talk about pleasure and pain. That's really what it's all about, isn't it?

A lot of human endeavor is focused squarely on chasing pleasure or avoiding pain, twin strategies that together provide the motivation for nearly everything we do, from eating and drinking to taking medicine and recreational drugs to enjoying music and movies to, well, fornicating. Pleasure and pain exist on a kind of continuum (more on that later), rewarding or punishing life's most ancient impulses to move away from what's bad and toward what's good.

Our sensations of pleasure and pain are not accidental. It isn't just dumb luck that we love the taste of bread and apples but not sewage or gasoline. Rather, we're living in the aftermath of millions of years of evolution selecting individuals for survival who preferred helpful things and spurned harmful ones. Pain hurts because a gene pool that develops a system that drives its creatures to quickly remove themselves from harmful situations outcompetes those that don't. Likewise, our clever brain provides us with a chemical reward, at the heart of what we call pleasure, for things that tend to preserve the species over the long haul. Sugar and honey taste good because they're good for us—they provide instant, usable energy. Orgasms rock because without them, sex would be nothing but awkward naked

MIND BENDERS

2 Hundredths of a second it takes for a human to register a photo of a person in a swimsuit as desirable or undesirable

3 Number of marijuana dispensaries for every Starbucks in Colorado

4 Years of childhood music lessons that improve brain functioning even 40 years later

65% of American singles have had sex with a coworker, according to Match.com.

wrestling, and our ancestors wouldn't have bothered to sire each next generation.

Interestingly, the brain uses the same basic mechanism for delivering chemical pleasure, no matter the source: dopamine. When neurons become active in a brain region called the ventral tegmental area, they trigger the release of sweet, sweet dopamine into the nucleus accumbens, part of the basal ganglia, a few inches behind each of your eyes. It doesn't matter what the trigger is—delicious chocolate, a satisfying workout, a quickie—do anything the brain deems beneficial, and it will give you the same cookie.

At the opposite extreme, pain is a little more complex—it's how our brains punish us for activities that might put survival at risk, like staring directly at the sun or sticking a fork in your ear. The triggering event—your hand on the hot stove, say—is often addressed with a reflex action, handled by the neurons of the spinal cord, that in this case helpfully pulls your hand away before you're aware of what's happening. The emotional aftermath (the pain) can go on much longer than the danger; that serves to strengthen that negative memory so you're more likely to adjust behavior and avoid the triggering event next time.

Further complicating the issue, pain often comes with its own remedy in the form of **endorphins**—homemade morphine-like drugs cooked up primarily by your pituitary gland that lower your perception of pain and can themselves produce a pleasant euphoria. If you like the burn of hot chili peppers, for example, it's not the pain itself, but the subsequent endorphin rush you want, and chasing that rush is partly behind why some of us gravitate toward things that hurt or disgust others, like roller coasters or Nickelback music. It may be a uniquely human trait to seek the pleasure in pain: Researchers have been unable to train rats to like chili peppers, for example.

This is the system evolution has bequeathed

Migraine sufferers have an unusually high libido—
20% higher than regular headache sufferers.

us. But being the clever apes that we are, we've learned many ways to hijack this natural reward/punishment circuitry, using the great influence we have over our brains' various inputs to trick them into delivering delicious chemical rewards we haven't earned. We indulge in recreational and medicinal drugs; we stimulate wakefulness with coffee and we wind down with a glass of wine after work; we seek sex purely for the orgasms and skip out on having babies. We're hacking the reward system to give us the payoffs without the costs.

It's all quite decadent, really.

And that's just the beginning of the ways we can influence our brain's pain/pleasure system: some conscious, some unconscious. We stimulate our brain's pleasure centers through exercise, through music, through meditation; we artificially surf naturally negative emotions and their endorphin "silver linings" with thrill rides and horror movies; we voluntarily bring ourselves to laughter and tears with comedy clubs and dramatic Netflix series and stirring Facebook posts about heroic dogs.

This chapter takes a closer look at some of the many, many ways we play with our brains, adjusting their inputs to score a sweet chemical fix. Sex, drugs, and rock & roll is only the beginning.

But let's start with sex, because why not?

There's arguably nothing more important than sex to any forward-thinking species, so it's no surprise the act involves a lot of input—in our case, from 30 or more areas of the brain. From arousal to orgasm, everything from trust to threat assessment has to be evaluated in preparation for the full-on sensory overload that happens if you're doing it right. Ancient brain regions like the amygdala let you access powerful emotions; the **anterior insula**, responsible for self-awareness, tells you you're feeling light-headed and have butterflies in your stomach; the hippocampus, which controls memory access, reminds you what goes where, and what name to shout out at the critical moment. The somatosensory cortex reports touch sensations from all over; it floods

"I'll have what *she's* having."

–Onlooker to Meg Ryan's public "orgasm,"
When Harry Met Sally

the brain with oxygen and inhibits pain receptors. Basically, it's a big old wonderful mess.

Are Orgasms Different for Men and Women?

At the moment of orgasm, the brain looks similar to the way it looks for people experiencing a heroin rush—and the contractions that characterize it look about the same for both genders. In fact, researchers evaluating people's descriptions of their own orgasms have trouble guessing the gender when anatomical references are removed. Orgasms, regulated by the limbic system, release two chemicals responsible for most of our sexual pleasure. Dopamine is responsible for lust fulfillment and the animal pleasure side of things. **Oxytocin**, often referred to as "the cuddle hormone," peaks at orgasm and creates feelings of warmth and bonding, of security and trust. It's released in women giving birth and in breastfeeding, too, and it makes you feel more attached to your mate, whether or not that mate is right for you. "Great if he's a good mate choice, but maybe not so much if you're Rihanna and he's Chris Brown," warns Catherine Salmon, Associate Professor of Psychology at the University of Redlands in an AlterNet.org article. "He may be a good lover but poor dad material, and yet you'll be attached to him and perhaps stick with the relationship longer than you should."

About a third of women have difficulty climaxing from penetrative sex. At the other extreme, about 15% can have multiple orgasms. Is there a difference between a "clitoral" and "vaginal" orgasm in women? According to recent research on this age-old debate, yes. In fact, scientists at Canada's Concordia University say that women can experience a variety of orgasms, even from the deft stimulation of non-genital areas like the nipples, neck, fingers, and toes. Nipple stimulation, for example, engages the same brain region, the medial paracentral lobule, that's engaged by clitoral, vaginal,

and cervical stimulation. The same region, by the way, is activated in men's brains.

Unlike men, whose genital tract has a single nerve pathway to the brain, women's genital tract has four. Rutgers University psychologists found that three of these nerve structures, in particular, seem to be involved in vaginal and clitoral pleasure: the pudendal nerve, responsible for relaying clitoral sensation to the brain, and the vagus and pelvic nerves, both responsible for vaginal and cervical sensation. The pudendal and pelvic nerves are connected to the spinal cord, but the vagus nerve bypasses it, which allows even women with damaged genital-to-brain nerve pathways from spinal cord injuries to feel vaginal and cervical pleasure and even climax.

*"As you begin to turn it on, the bulb begins to get bright, then brighter, and brighter and **finally in a blinding flash is fully lit.**"*

–Marilyn Monroe, describing her orgasms

For men, ejaculation can't happen again right away, because the body requires a resting "refractory" period of 30 minutes to 24 hours, which increases with age. This leads many men to assume they can't have multiple orgasms, but this turns out to be false. While men can't ejaculate again and again without rest, through careful training and practice they CAN learn to *orgasm without ejaculating*, allowing them to share in the fun of multiple orgasms. Adult education we can all get behind.

How Do We Choose a Mate?

What happens in that millisecond when you see someone *hot*?

We've all seen a photo mosaic, when a bunch of small pictures combine to make a bigger one. Our brains compute the image in two ways: by "global" processing, which helps us see the "big" picture, and by "local" processing, which helps us see the smaller ones. These cognitive systems help us spot a hottie, too.

Psychologists at University of Nebraska-Lincoln found that women's brains take the global route when they see a cute guy, focusing on his whole being rather than individual physical features. Interestingly, men see other men this way, too. But, men's brains take the local route when they see a cute woman,

viewing her as the sum of her body parts. Local processing is how we see objects, so in this case, the scientists explain, women are literally objectified.

What if you insist that you pay more attention to a woman's personality than her looks? Here's a reality check: Dr. Sarah Gervais of the University of Nebraska-Lincoln led a follow-up study using eye-tracking technology to trace men's and women's gazes. Take a guess at what she found. Guess your fiancée really did ogle that waitress after all.

Surprisingly, women view *each other* that same way. This could be, in theory, an evolutionary response to scope out mating competition, or perhaps a result of our hyper-masculine media and pop culture socializing them to adopt the male gaze.

But that's not the only stuff happening. The orbitofrontal cortex activates upon sight of attractive faces. In heterosexual men, the amygdala has a visceral reaction to both highly attractive and highly *un*attractive female faces.

Additionally, two sub-sections of the medial prefrontal cortex help determine who's worth a second look. The paracingulate cortex activates upon the sight of someone who's considered conventionally attractive—a *Sports Illustrated* cover girl, for example. The rostro-medial prefrontal cortex, on the other hand, engages upon seeing someone you, personally, are attracted to. The latter plays a crucial role in social decisions and helps you evaluate whether someone's a good match *for you*. Attraction: It's all in your head.

What goes on in our minds during courtship? According to Dr. Gad Saad in *Psychology Today*, people in the market for short-term hookups often choose based on "noncompensatory" models, like "I want to have sex tonight with somebody who is at least a '7.'" But for long-term mating, they tend to use "compensatory" models that weigh various attributes such as looks, status, or kindness, according to personal preference. In this kind of behavioral modeling, even a man or woman for whom looks is the most important element can be swayed to a less beautiful mate choice who excels in multiple other, less-important areas. Additionally, according to a 2009 study reported in *Psychology Today*, women reject potential suitors more quickly than men, and they consider more options before choosing a mate. This accords well with the general notion of "parental investment" in nature, where the parent likely to make the biggest investment in raising children from the union is more judicious in choosing a partner.

How Fetishes Happen

How do you go from being a normal, curious sexual novice to being sexually attracted to sweaty footwear or people dressed as hamsters? It's one of the most fascinating areas of human sexuality. One possible explanation for the emergence of fetishes is **neural crosstalk**. The somatosensory cortex regions for feet and genitalia are adjacent, for example, and leaky messaging could result in foot fetishes—among the most commonly reported of all fetishes.

Another answer is conditioning. Like those famous "Pavlov's dogs" that learned to salivate at the ring of a bell after associating it with food rewards, if you're repeatedly exposed to a random trigger while sexually aroused, the trigger can begin to cause arousal. In one study from the '60s, men who were shown naked women and boots enough times were attracted to the boots even after the ladies were taken away—which led to a nice hit record for Nancy Sinatra. Pleasure and pain, involved in the same chemical reward system in the brain, can become commingled through association: If the use of handcuffs during sex play turns you on, perhaps the site of a police cruiser could eventually have the same effect.

What about the erotic associations some have around

even kinkier triggers like bodily fluids, violence, and so on? Sexual arousal is obviously critical to the survival of your genes, and a lot of the usual rules get pushed aside in its service. Your sense of disgust lessens when sexually aroused, so getting turned on by *My Little Pony: Friendship is Magic*, for example, can feel more "normal" than it ordinarily would, allowing it to become, through conditioning, a kink that works for you.

In particular, sadomasochism, it's been found, can lead to an out-of-body state of consciousness like that sometimes attained through yoga or meditation. Practitioners can be in a state where more **cortisol** than usual floods the brain, indicating stress, and yet they report *less* stress than a control group. Researchers surmise it's because the activity, particularly for those on the receiving end of the pain, can cause blood to flow away from areas of the brain linked to executive control and the ability to distinguishing "self" from "other," leading to an experience where they're divorced from their own stress.

Why Is There No Female Version of Viagra?

When it comes to sexual dysfunction and low libido, men have a little blue option. Women with low libido have historically been on their own, but thanks to flibanserin, a drug made by Sprout Pharmaceuticals and marketed under the trade name Addyi, it may be Ladies' Night at last. After a rocky start, Addyi's been approved by the FDA to treat a condition known as hypoactive sexual desire disorder (HSDD), otherwise known as low libido. "From my neck down, my body responds perfectly," bemoaned Amanda Parrish, a participant in flibanserin's clinical trials, in *Time* magazine. "What's missing is the lack of desire to start."

Addressing low female libido is a more complex proposition than the blood-flow-to-the-popsicle issue Viagra addresses in men. Flibanserin works its Barry White-like magic by modulating various sex chemicals in the frontal cortex, increasing some (like

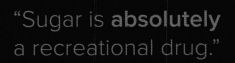

Figure A

Figure B

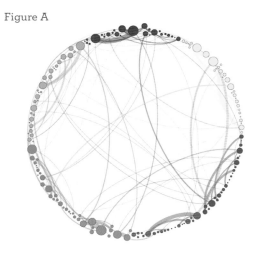

dopamine) and decreasing others (like serotonin). Similar to Viagra, which was originally a cardiovascular drug until somebody smelled the profit in the side effect of a four-hour erection, flibanserin was created to treat depression before being politely repositioned. Side effects can include nausea, low blood pressure, and fainting, and testing demonstrates only a modest improvement in horniness. But try telling that to Parrish. "I know this pill worked for me," she claimed in June of 2015, after her clinical trial had ended and her love life had sadly ebbed again. "I hope that it's approved. I want to want my husband again."

What Drugs Do

"Drugs" is the umbrella term for the ingested substances we use to hijack our brain's chem-

ical reward system and access different conscious experiences—and it's big, big business. The global revenue of the pharmaceutical market tops a trillion dollars a year, now, and another estimated $430 billion is spent on illegal drug trafficking. Here's some helpful detail around what a number of drugs—both legal *and* illegal—do inside your head.

Over the Counter

ALCOHOL

No matter how loud and obnoxiously you sing karaoke, alcohol is a depressant, chemically speaking. In your brain, it blocks and supports different neurotransmitters, lowering your inhibitions and often changing the course of the night, for better or for worse. It also numbs pain and makes you sleepy. Alcohol is extremely

"I used to have a drug problem. **But now I make enough money.**"

—David Lee Roth, Rock star

The sober spider

Marijuana

Chloral hydrate (a heavy sedative)

Benzedrine (an amphetamine)

A DAY IN THE LIFE
OF YOUR BRAIN

You don't have to be tripping on krokodil or fairy dust to give your brain a ride—your normal day is an up-and-down roller coaster of chemical stimulation. Here's what's going on.

MORNING COFFEE

You don't really come alive, as a card-carrying adult, until your brain is tricked into wakefulness by caffeine molecules. Here's how it happens: Caffeine blocks the neurotransmitter adenosine from building up throughout the day to make you feel tired. Result: you don't know how exhausted you are.

EXERCISE

A cardiovascular workout in the morning provides a rush of endorphins that enhance your mood, soothe anxiety, and improve memory function. A regular exercise routine reduces stress, fights anxiety and depression, and grows your memory-processing hippocampus. And no slacking off—research suggests that you'll start to lose the brain benefits after just one week of inactivity.

AFTER WORK DRINKS

Depending on what you're drinking, you're flooding your brain with alcohol, sugar, and caffeine (looking at you, Rum & Coke). The ethanol in the alcohol interferes with neurotransmitter activity, loosening your inhibitions, slowing reaction time, and making Lyft an increasingly smart option.

DINNER

A family dinner confers a host of emotional benefits, especially for kids. Kids whose families eat together five times or more a week have better vocabularies, are less likely to drink or use drugs, are better adjusted emotionally, and do better in school. But keep the TV and all other screens off: The magic doesn't work if you're sitting in the same room but not communicating.

SEXY FUN TIME

At climax, activity spikes across the brain; after orgasm, the brain releases oxytocin and dopamine, producing feelings of bonding (ooh) and pleasure (aah).

SLEEP

When your exhausted brain works all that caffeine out of its system, the normal adenosine receptors will tell your body it's time to crash.

BREAKFAST

A balanced meal of complex carbohydrates and protein, such as toast and eggs, boosts neurotransmitters and improves learning ability. Choose wisely: Blueberries can stave off cognitive decline and dementia, and avocados can promote healthier blood flow in the brain, but the sugar in donuts inflames and bogs down your hippocampus.

WORK

Boring meetings, missed deadlines, mercurial bosses: Work stress produces the chemical cortisol; over time, this can strengthen connections between the amygdala and the hypothalamus, keeping the mind in a constant state of fight-or-flight, fear and aggression, and make you chronically anxious and depressed. Work can be essential to well-being and mental stimulation—but keep your perspective.

LUNCH

As with breakfast, it's all about choices. Kale and other cruciferous vegetables have powerful antioxidants; a salad can root out damaging free radicals. A diet rich in fried food, on the other hand, can build plaque in your brain and is associated with smaller brain size.

AFTERNOON SNICKERS BREAK

Despite the sugar, chocolate has an unexpectedly positive effect on brain function. Like coffee and tea, chocolate has **methylxanthines**, which improve concentration, and nutrients called "cocoa flavanols" which boost blood flow to the brain and improve working memory. Did you need an excuse for an afternoon chocolate bomb? You're welcome.

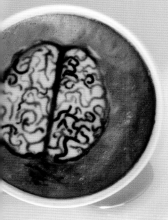

**NO IFS, ANDS,
OR BUTTS**
The brains of
smokers produce
less dopamine than
non-smokers, but
can be reversed
by quitting.

addictive, and binging or long-
term abuse can actually reduce the
size of brain cells, shrinking your overall brain
mass and leading to side effects like behavioral changes, sleep difficulty, mood swings,
and memory loss. In one study, habitual binge
drinkers took longer to come up with the
same solution to problems than their less-fun
counterparts.

SUGAR

Sugar is perhaps our very favorite addictive
substance, and the only one we actively pimp
to our children. (Looking at you, Easter Bunny.)
Under assumed names like dextrose, maltose,
lactose, and their morbidly obese cousin high
fructose corn syrup, sugars are the oil barrels
of the natural world, offering compact packs of
easily accessible energy. When you're eating
sugar, your brain lights up in the same areas
as a binging alcoholic, and it releases dopamine for pleasure and serotonin for calmness.
(The myth of the sugar high has been widely
debunked; turns out kids are just excited about
cookies.) With a whole box of Krispy Kremes
your insulin shoots up and then drops, leaving you craving more. Evolution simply hasn't
prepared us for a world where this rare, prized
commodity is so readily available, and artificial
sweeteners may be worse—they trigger the
desire but not the reward.

TOBACCO

Given the obvious health risks around
ingesting nicotine and tar, and the mildness
of its high relative to almost everything else,
it seems a wonder it's as popular as it is—until
you realize the tobacco companies long ago
began manipulating nicotine levels to make
it more addictive. On its own, tobacco stimulates your mood and suppresses appetite. Its
active ingredient, nicotine, is absorbed quickly
into your bloodstream and travels to the brain,
where it causes the release of adrenaline. You
feel energized, maybe lightheaded; when the
buzz fades (after as soon as 15 minutes) you
feel a little down and tired and sure could use
another cigarette, if anyone's offering. Long-
term use can weaken your senses of taste and
smell, and, yeah: kill you dead.

CAFFEINE

Available in dosages like Grande and Venti,
caffeine is in almost universal use among busy
adults. Interestingly, at the chemical level caffeine doesn't give you a boost at all—rather, it
blocks a system that tries to make you sleepy
when your brain has been working too hard.
The firing of neurons in your active, waking
brain produces a chemical called adenosine.

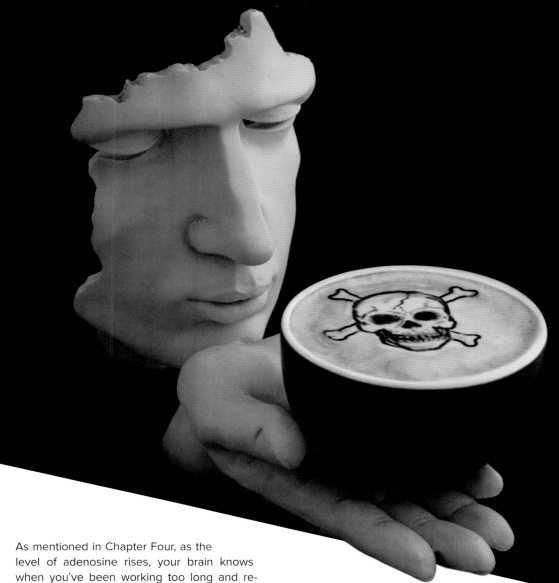

As mentioned in Chapter Four, as the level of adenosine rises, your brain knows when you've been working too long and releases drowsy signals to seduce you into slowing down. Caffeine gets all up in there and blocks those receptors, so you never get the "we're done here" message. You can build up a tolerance to a certain level of caffeine after a week to 12 days, after which you need more to get the stimulant effect. When people come out of surgery and describe themselves as having a headache, it's often just caffeine withdrawal.

Under the Counter

MARIJUANA

The active ingredient in weed, grandly called "delta-9 tetrahydracannabinol" (or by its rapper name Run-THC), overwhelms a brain system, retroactively called the **endocannabinoid system**, and acts as a sort of dimmer switch, slowing down synaptic firings between neurons all over your brain. THC delivers various effects, depending on which receptors it's blocking in this system. When THC gums up the circuits of your emotion and fear-regulating amygdala, it produces the infamous paranoia. THC in your responsible-for-new-learning hippocampus makes it hard for you to follow a conversation, and THC in your autonomic nervous response brainstem delivers anti-nausea effects. It's important to note, however, that unlike the other substances addressed in this chapter, marijuana is *not* chemically addictive. More on that later.

COCAINE

Cocaine, the powderized extract of leaves from the coca plant, overwhelms the brain with dopamine and other neurotransmitters by

STRESSED?
If you're already stressed, caffeine is not your friend—it both increases your body's stress response (raises heart rate and blood pressure) and alters your perception so the stress feels worse. Try a stress ball instead...

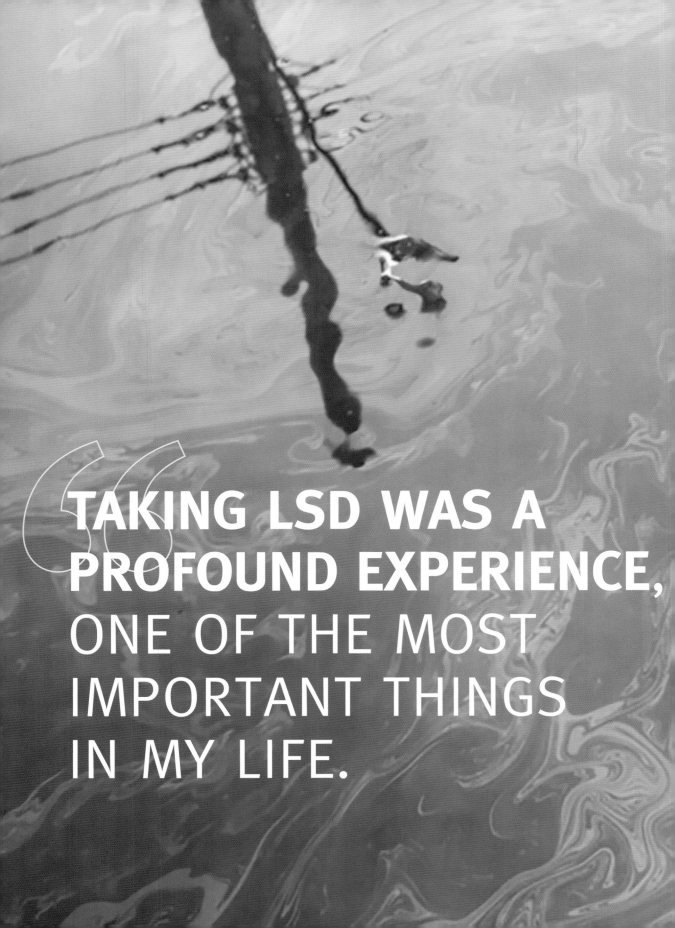

"TAKING LSD WAS A PROFOUND EXPERIENCE, ONE OF THE MOST IMPORTANT THINGS IN MY LIFE.

LSD SHOWS YOU THAT THERE'S ANOTHER SIDE TO THE COIN, AND YOU CAN'T REMEMBER IT WHEN IT WEARS OFF, **BUT YOU KNOW IT."**

–Steve Jobs

preventing their natural reabsorption. Whether in snortable or smokable form, it's extremely addictive (though not after a single use, as is popularly alleged), to the point where the requirement to get more cocaine can overwhelm natural instincts: lab mice, like crack-addicted humans, have been shown to prefer cocaine to a comfortable environment.

ECSTACY

12 million Americans or more have tried ecstasy, a.k.a. MDMA, "molly," and the "Love drug." Popular at rave parties, ecstasy's reported effects include increased physical and emotional sensitivity and heightened awareness of color and music. In the brain, it's been described as "opening the floodgates" for neurotransmitters, like dopamine, norepinephrine, and mostly serotonin, resulting in the elevated mood. Ecstasy decreases your awareness of negative and threatening stimuli, resulting in a positive bias, of sorts, in which you feel loving and socially connected. Taking ecstasy, according to studies, can affect your memory and ability to perform simple tasks. It can also be risky—MDMA use can bring on dehydration and impair your ability to regulate your body temperature, which have led to a lot of ER visits and some deaths.

MAGIC MUSHROOMS

These **psychoactive** fungi that grow naturally on horse droppings and in other disgusting places were long a natural favorite of Native Americans and fans of The Doors. The active drug, psilocybin, facilitates connections between parts of the brain that normally don't communicate, resulting in "trippy" thoughts and

> # "Music gives a soul to the universe, wings to the mind, flight to the imagination, and life to everything."
>
> *Attributed, through liberal translation, to* **Plato**

hallucinations, and an altered, more open mind-set that some report can last a year or more.

LSD/ACID

Acid, or lysergic acid diethylamide (LSD), is a psychedelic substance synthesized by Dr. Albert Hofmann in 1938, though it didn't become a recreational juggernaut until the '60s. What literally happens in your head, chemically, can't begin to explain the subjective experience, which is famously profound and hard to describe to LSD-virgins. The easy part: In your head, the compound affects the prefrontal cortex, home to mood, cognition, and perception. All hallucinogens temporarily disrupt your natural neural pathways and replace them with new connections; your brain, uncomfortable with randomness, tries to make sense of these. LSD in particular, according to one recent study, flips on switches that are normally activated by the external senses—this means that wholly internal thoughts register as external sights, sounds, etc., and these hallucinatory effects can last for hours.

What Is Addiction?

For most of humankind's long, sad history of trying to deal with substance abuse, society has treated addiction as moral failing worthy of punishment. Only recently have we started to rethink addiction as a chronic disease, or a bad habit out of control, worthy of humane treatment. One reason it's easy to blame the victim is that, unlike with some chronic diseases, drug addiction typically begins with voluntary acts, like smoking cigarettes, or shotgunning that first beer behind your junior high. But make no mistake, addiction isn't a long series of bad choices—an addict is a person only partially in control of a brain rewired to prize the addictive substance above everything else.

How does this happen? Well, dopamine, which delivers the pleasure that made you want the experience in the first place, is also involved in memory and learning. Current theory posits that, in addicts, dopamine interacts with glutamate (another neurotransmitter) and overloads a natural reward chain that involves not just pleasure but motivation and memory. In other words, we transition from "I like this" to "I need this, now and on an ongoing basis, even if I have to sell my grandma's toaster to get more."

We had a sort of wrongheaded idea for decades that there was a meaningful distinction between physical or psychological addictions, but it no longer seems a very important distinction. An addiction can be defined as an irre-

> **CAN'T GET A SONG OUT OF YOUR HEAD?** Chew gum! It interferes with your "articulatory motor programming" process, which governs verbal memory and seems to be a culprit in replaying songs in your head.

Music helps you exercise: Cyclists who listen to music consume 7% less oxygen than their counterparts riding in silence.

sistible craving that persists even in the face of severe physical and social consequences. Meth addicts know what's happening to their faces, teeth, and overall health and well-being; they just can't stop the downward spiral by force of will.

Why Is Addiction So Hard to Cure?

Addictive drugs cause the release of increased amounts of neurotransmitters, and the brain copes with the excess in a number of ways, some of which have long-term effects. They can reduce the number of dopamine receptors, making you need more of the drugs to get you high next time. This is called tolerance, and there's a catch: As you become less responsive to this drug, you can also become less responsive to other things that used to make you happy, thus distorting cognitive and emotional functions and starting to hardwire habitual behaviors. Now you're an addict. Many former addicts say addiction represents a one-way threshold: Your brain is in some ways permanently rewired, and it's hard or impossible to go back to occasional use. One reason: A return to your old social and environmental cues can easily trigger the old cravings

and lead to relapses.

Addiction is difficult to treat, never mind cure, but one promising avenue is anti-drug vaccines. These stop the drug from entering your brain, so you don't experience the high: no dopamine-driven pleasure, no addiction. Well-designed antibodies attach to the drug molecules in your bloodstream, tagging them so that white blood cells can find and destroy them as invaders. This innovative approach has so far been successful in mice and monkeys; researchers have already begun human trials.

This Is Your Brain on Rock & Roll

Music has been part of every human culture ever found. We listen to music while we drive, work out, eat, and have sex; we spend hundreds to watch live musical performances; we soundtrack our weddings and funerals and everything in between. There's something primal about it: Babies respond to rhythm before they can speak. What does music do inside our brain that's worth this tremendous investment?

Music processing takes place mostly in the auditory

What's the deal with the "Runner's High?"

Is it real, or just a sneaky plot to trick normal people into exercising?

This exercise-induced euphoria appears to be the result of endo-cannabinoids and/or endorphins, morphine-like brain chemicals that the body produces in response to pain and discomfort. One study showed that after a two-hour run, runners' endorphin levels correlated well with their self-reported euphoria. Researchers found that the same parts of the brain that respond to love light up during this feel-good state. If what you're chasing is better marathon numbers, the key is to push yourself as hard as you can without going over the edge into exhaustion or injury. For more tips on how to achieve this high, run over to page 144.

FINE-TUNE
YOUR WORKOUT

Check out these workout activities by age group—and samples of songs you'll know that are close in terms of target beats per minute. (Check with your doctor to make sure working out is right for you.)

ACTIVITY & AGE	TARGET HEART RATE	SAMPLE SONG AT THIS BPM*
STRETCHING		
At 25	115 BPM	"Uptown Funk," Mark Ronson, Bruno Mars
At 35	108 BPM	"Just a Girl," No Doubt
At 45	92 BPM	"Give it Away," Red Hot Chili Peppers
WEIGHTLIFTING		
At 25	130 BPM	"Elastic Heart," Sia
At 35	129 BPM	"I'm Shipping Up to Boston," Dropkick Murphys
At 45	110 BPM	"Another One Bites the Dust," Queen
BICYCLING		
At 25	149 BPM	"Bang Bang (feat. Ariana Grande, Nicki Minaj)," Jessie J
At 35	134 BPM	"Material Girl," Madonna
At 45	133 BPM	"Superfreak," Rick James
SWIMMING		
At 25	160 BPM	"Happy," Pharrell
At 35	146 BPM	"Can't Hold Us (feat. Ray Dalton)," Macklemore & Ryan Lewis
At 45	134 BPM	"Fight for Your Right," Beastie Boys
SPRINTING		
At 25	186 BPM	"American Idiot," Green Day
At 35	180 BPM	"Eye of the Tiger," Survivor
At 45	163 BPM	"Paint It Black," The Rolling Stones

*Beats per minute

ing in a choral group releases stress, and that moving together with a crowd (doing the Macarena or the "Nay Nay," say) leads us to subconsciously feel at one with others in multiple ways, including predicting they'll share our opinions on unrelated matters. Participating in a rhythm makes people feel part of a cohesive group, which is a good survival tactic for any forward-thinking society.

While music has helped society along, it's also good for us at a personal level. Ambient noise played at a moderate volume can improve creativity, for example. Does it interrupt your train of thought? Yes, that's the point; it increases processing difficulty, and leads your brain to abstract processing, which is one key to unlocking creativity. And music may even affect how selfish or generous you are: A Japanese study found that participants who listened to music they liked donated more money than they did in a silent control scenario. When listening to music they disliked, they donated less.

As Friedrich Nietzsche puts it, "Without music, life would be a mistake."

Why Do We Love Drumming?

The grandmother of longtime Grateful Dead drummer Mickey Hart was suffering from dementia. But when he played drums for her, she said his name—for the first time in a year or more—inspiring him to begin a long investigation into music as therapy. Today, when you go to a Mickey Hart Band concert, you'll see an unusual light show: He's hooked up a cap that records EEG information from his brain waves, and—thanks to super-fast processors—powers the light show at concerts. You're watching his brain in real time.

Drums are probably the oldest instrument in the world. Stretched animal skins have helped

If your go-to masseuse is booked until the next millennium, listen to "Weightless" by Marconi Union. Scientifically, it has been deemed the "world's most relaxing song" by Mindlab International.

cortex, right at the top of your temporal lobes—to find them, run a finger from your temples straight back like you're defining a terrible bowl haircut. The music experience is deeply rooted in the ancient limbic system, and music affects our emotions instantly. Testing shows that if you play a sad or happy song to people, then they are more likely to interpret a neutral face as sad or happy, depending on which type of music they heard.

Why does music affect us at this emotional level? One theory is that music, which predates and transcends human language, evolved socially as a way to convey calm or excitement or fear—it's our version of animal calls. Music may have helped glue preliterate societies together. Neuroscience research suggests that sing-us keep the beat since prehistoric times, all over the world. They've powered shamanistic drum circles, set the pace for foot-soldier marching, and synchronized Viking ship rowing. What is it about the drumbeat that's so compelling and universal?

The connection may have something to do with our heartbeats. Your heartbeat rate is not fixed: It changes with exercise and rest, and you can learn to adjust it through breathing and yoga techniques; fetuses can synchronize their heartbeats with their mother, and choir singers with one another.

Whether it's intentional or accidental, the drumbeats underlying popular songs are more or less equivalent to the heartbeat you'd experience in whatever emotional state or activity the songs evoke. A normal resting heart rate is 60-100 beats per minute, corresponding to chill tunes like Childish Gambino's "Redbone" (80 BPM) or The Beatles' "Come Together" (82 BPM). Dance music's beat— like Shakira's "Hips Don't Lie" (100 BPM) or Michael Jackson's "Don't Stop 'Til You Get Enough." (119 BPM)—is closer to the beat of your heart during physical activities. It may be that the music that gets you dancing does so by convincing your brain that you're *already* dancing.

Music To Work Out By

Music conveys a number of brain benefits for people working out. The beat can help you stay in rhythm and avoid the false steps that cost you energy and disrupt your breathing. Your body sends specific physiological feedback to your brain, like skin temperature and the level of lactic acid in your muscles, that the brain calculates; if it thinks you're overdoing it, it initiates shutdown procedures (which, to you, feels like you're exhausted). Music can distract you from this, letting you dig deeper and push harder.

But there's more. Your heartbeat rises in response to exercise, and it also can change in response to the music you're listening to. If you can align your workout tunes with your target heartbeat, it can help you get a more efficient workout. See the chart on page 142 to begin crafting your neuro-perfect workout playlist.

What's the "Runner's High" All About?

Long distance runners have a secret motivation beyond those teeny paper cups of Gatorade bystanders hand out: a mysterious "runner's high" that reduces pain and provides a mild euphoria. The effect appears to be the result of endocannabinoids and/or endorphins, our bodies' morphine-like brain chemicals produced in response to pain and discomfort. It's thought the "runner's high" helped our ancestors ignore body pain while chasing down prey or escaping predators—even if what you're chasing is better marathon numbers, the key is to push yourself hard physically without going over the edge into exhaustion or injury.

Not everyone experiences the "runner's high," and not everyone who does experiences it every time. For your best shot at seeing what this feels like, according to the pros, try a) getting eight hours of sleep, b) running in the morning, c) running with friends and to music, and d) being just a little stressed when you start—all of these things should improve your odds.

Humans are self-aware at a very thorough (and ever-increasing) level. Alone in the animal kingdom, we willingly manipulate our brain's inputs to tap into our brains' reward circuits for pure recreational pleasure, or to try expanding our mental experience. But to what end? What are we striving for, exactly?

In the next chapter, we'll focus squarely on some of our minds' highest-order functions: emotions, faith, spirituality, and what it means to believe.

Music could be thought of as **"a type of legal performance-enhancing drug"** according to Costas Karageorghis, one of the world's leading experts on the psychology of exercise music.

CHAPTER 8

EMOTION, FAITH, SPIRITUALITY, AND THE RISE OF CULTS

Intelligence is important, but when it comes to decision-making and thought processing, there's more to you than just your cerebral cortex.

Human thought is characterized by a complex interplay between the rational and the emotional, between what you think and what you feel. Our vastly expanded cerebral cortex opened the doorway to so-called higher-order thinking like philosophy and art and reason. But that higher-order thinking didn't replace the reactionary animal intelligence that came before—it only supplemented it.

Rational thinking and emotional feeling aren't different flavors of the same mental process—they operate in totally different systems within the brain. Your emotional reactions spring from tightly choreographed synchronization among structures in the older limbic system, consisting primarily of the hypothalamus, thalamus, amygdala, and hippocampus, and sometimes called the brain's emotional center. This is the hot and ancient core of your brain, through which sensory perceptions, memories, nervous responses, and anxiety pass. Its activity is tied closely to automatic responses and chemical messengers beyond your active control. The push and pull between the reactions of this

MIND BENDERS

35,000

BCE: Oldest of mysterious "Venus figurines" found in gravesites across Europe

2,100

BCE: Creation of Gilgamesh, humanity's oldest surviving literary work

4,200

Estimated number of different religions in existence today

In 2016, 89% of Americans said they believe in God...up three points from the 86% who said that in 2007.

emotional center and the musings of the cerebral cortex essentially forms the basis for your opinions and beliefs.

The interplay between these two systems leads to a host of distinctly human behaviors. Consider situations like these: how you'll allow a gut instinct to override a rational assessment; why we stay with friends and lovers we know are bad for us; how important it is to work with people we trust; how frustration, lust, or anger can overturn the most carefully considered plan. Ask for advice on a tough decision and you'll nearly always receive a vague recommendation to strike a balance between the two: Use your head, but trust your heart. When asked what you believe, you will likely find yourself actively combining elements of the rational and the emotional.

As fascinating as belief is at the individual level, it gets much more interesting when a large number of individuals share a particular belief because the group effect is a potent force multiplier. For the vast unprovable parts of anything I believe, it helps firm my resolve to surround myself with others that feel the same way. Thus preferences become trends; belief patterns become morality and ideology; faith becomes religion; similarity becomes community; disagreement becomes war. And in the same way that the cerebral cortex is subject to all manner of ailments and conditions like Alzheimer's disease and strokes, our emotional limbic system is not immune from breakdowns and hijacking, leading us to anomalies like alternative facts, groupthink, mob behavior, and, yes, cults.

So let's investigate the mysterious world of what we can't know for sure—that infinitely vast universe of truths that are immune to our senses, beyond our intellect, and possibly even outside our grasp. Let's talk about what we can know but never prove.

How Emotions Happen

We often use the terms **emotions** and **feelings** interchangeably, but to neuroscientists, emotions are your body's automatic responses to outside stimuli; feelings are your brain's interpretations of those responses. On ScientificAmerican.com, neurologist Antonio Damasio explains: An outside stimulus (say, a raccoon in your garage) triggers your heart to

Turns out we really are blinded by jealousy. A University of Delaware study found that jealous women failed to notice images they were tasked to find, and the more jealous they were, the worse they did.

race and your mouth to dry up, as your body prepares for fight or flight. Those are emotions—involuntary and unfiltered, the byproduct of a body that evolved to respond quickly, long before we brought our fancy expanded cerebral cortices into play. Next the limbic system, which governs fear, aggression, and so on, interprets those physical reactions and categorizes them—in this case, as the feeling we call fear. Finally, your self-awareness creates a coherent story out of this categorized natural reaction and decides you're afraid but resolved, and should go get your raccoon broom.

The source of those purely physical reactions is the autonomic nervous system (ANS), which is centered in the brain stem at the base of your skull, but has neural branches stretching all over the body. The ANS governs a lot of unconscious body behavior, from heartbeats to sweating to coughing to being turned on sexually. It's connected to your limbic system by two branches: the **sympathetic** (which excites you—tells your eyes to dilate, your heart to race, and so on) and the **parasympathetic** (which calms you down once the crisis is past). In between, the various elements of the limbic system try to figure out what's going on, and what to do.

The amygdalae—as a reminder, each hemisphere has one—seem to regulate fear and threat response, for example, and may help process information about positive and negative events. Researchers have shown that the left and right amygdalae show separate and distinct roles in the processing of emotion, with damage to one side or the other having different impacts on our mental faculties. Damage to the left amygdala can increase depression, while damage to the right can greatly reduce a subject's arousal response to pain, for example. These circuits tell us how to feel; it may be that some mental conditions arise as the result of the brain failing to correctly attribute the usual emotional response to an event. Depression, in this paradigm, comes about when the brain simply no longer associates joy with an activity or experience that it previously did. You quite literally lose the ability to enjoy things that once made you happy.

Emotions Interpreted: What Feelings Are

So emotions are automatic responses; feelings are how we explain those responses to ourselves. We often think of feelings as mirror images of each other—happy/sad, proud/ashamed, optimistic/pessimistic. It might be easy to imagine they're related that way in the brain, where your experience of each corresponds to a point on a simple continuum. But that doesn't appear to be the case. Twenty years ago Dr. Mark George, a psychiatrist and neurologist with the National Institute of Mental Health, demonstrated that happy and sad feelings move along entirely independent paths. Looking at the left and right amygdalae, Dr. George found that sadness yielded an increase in amygdala activity on both sides, while happiness triggered a differential response: more activity in the right and less on the left. We can and do experience opposing feelings at once—like the bittersweet, simultaneous pull of happy and sad that parents experience when dropping a child off at college, for example.

Negative feelings can be debilitating, but they have their purpose, too. Take disgust, for example. If you loved the smell and taste of everything, visits to bad restaurants might be more enjoyable, but they'd be more dangerous. Disgust is in our genome precisely because it protected us, historically, from certain discernible dangers like poisonous foods and microbe-ridden filth. The estrogen-driven increase in sensitivity to noxious smells that many expectant mothers report may persist today because it helped protect unborn children throughout our history. And some research suggests that obsessive compulsive disorder, or OCD, may be a result of the brain's disgust center becoming hyperactive, registering disgust even for things most people don't have a problem with (like a disorganized sock

Humans have **43 facial muscles** that work together to display **over 100 distinct emotions.**

drawer), causing the sufferer to respond with rituals or behavior that provide comfort and a sense of order.

Memories that have emotional connections are stored and retrieved more easily, and remembered more vividly. This is because, inside the limbic system, the amygdalae and other feeling generators work closely with the memory-generating hippocampi. Our brains deem emotional responses important, and they've evolved to make sure you'll remember that the teacher yelled at you—*Warning! Survival threat!*—in front of the class—*Warning! Potential loss of social standing!*—long after you've forgotten whatever you were studying that day.

Why You're Hooked On a Feeling

Here's a quick look at what's happening—and why—when you experience some familiar feelings.

LOVE

Love is a little more complex than just an emotion or feeling. Rather, it's an attachment and state of mind, to which anyone who's ever been in love will attest. Love has stirred poets from William Shakespeare to Adele; some even say it makes the world go 'round. There are as many kinds of love as there are human relationships: the love of parent to child, friend to friend, acolyte to master, stranger to pop star. But they all share a certain magic—a desire for a deep connection, a hunger to join and be accepted, a hopefulness held up by joy and dragged down by nauseous despair, a physical rush in the brain and throughout the body.

Love is a many splendored thing. Love will keep us together, but love hurts; love stinks; love is a battlefield. Love is higher than a mountain; love is thicker than water—in fact, to listen to our artists, all you need is love.

Love brings with it some of the most striking and powerful physiological changes a body can experience, which is part of why it affects us so deeply. If your palms get sweaty and your heart races when you see the object of your secret crush, it's because the fear and stress of the situation activate your autonomic nervous system (ANS). And much the way it does during the more familiar fight-or-flight response, your ANS is preparing your body to take decisive, dangerous action. Like daring to ask her out. Or running away.

fMRIs show that love causes increased blood flow to the brain and lights up your pleasure center nucleus accumbens like Vegas, as love's three main components—lust, attraction, and attachment—flood the limbic region with actual drugs. Lust delivers a shot of adrenaline and norepinephrine that produces sweaty hands and an increased heart rate. Attraction activates your opioid system, giving you a morphine-like high. And attachment kicks up the hormones oxytocin and vasopressin, bringing a sense of well-being and safety.

You're gonna have to face it—you're addicted to love.

HAPPINESS

Sadly, happiness is not our natural state, according to neuroscientists. Humans have a natural "negativity bias," because until recently we were clawless, armor-less, mostly-naked animals trying to survive in a dangerous world, and animals that thought about the worst-case scenario first were more likely to see another day. If whitetail deer don't look happy, now you know why.

And even in our relatively cushy modern situation, with HDTV and at least 13 different shapes of french fries, neuropsychologists caution against trying to eliminate negative thoughts and recommend focusing instead on appreciating the positive ones more. Successful couples, according to research cited in *Psychology Today*, aren't those who never fight—they're the ones who manage to balance out each negative interaction with, on average, at least five positive interactions.

Given the negativity bias, why are there happy people? Part of the answer may be physiological: Researchers led by Dr. Wataru Sato of Kyoto University recently discovered that happiness correlates with a higher volume of gray matter in a particular area of the parietal lobe called the **precuneus**. It's an area involved in self-reflection and one known to be affected by meditation. And part of the answer may lie in our approach to life. In a 2011 article in the *Guardian*, psychologists theorized that happy people's dispositions stem from an enhanced ability to recognize and appreciate positive events—they create a "cycle of positivity" that does not include naïveté or an obliviousness to danger.

For the part of happiness you can control, at least, it helps to savor the good and not dwell on the bad.

GUILT

Hate yourself a little for cheating on your taxes, on your spouse, or on your spouse's taxes? The brain's anterior temporal lobe is key to understanding how our actions are interpreted in a social context: It's the home of guilt. When the anterior temporal lobe is working in harmony with the frontolimbic network, where strong emotions are controlled, you can wisely weigh your own behavior against the behavior of others and make reasonable decisions about what's acceptable. In some people with recurring depression, a recent fMRI study found these two normally communicating regions are disconnected—meaning the frontolimbic network has to make decisions about emotion without input from the context-providing anterior temporal lobe. Without the usual logic of social context and understanding, your brain can feel strong negative emotions but not know where to attribute them, and without a logical exterior target, you pin the blame on yourself—otherwise known as guilt.

JEALOUSY/ENVY

While you can get drunk on love, jealousy is your evil drinking buddy. *Psychology Today* summarizes that jealousy involves three people, whereas envy—often mistakenly used interchangeably—involves two people. Envy is the brain's reaction to wanting something that someone else has, whereas jealousy is the brain's reaction to the threat of losing something, or someone, to someone else.

A 2015 study conducted by Shenzhen University in China found that when we're jealous, the brain's empathy center, the anterior insular cortex, becomes suspiciously quiet. Study participants took a test, and were all told cruelly afterward that they were only "two-star" performers; later, while being monitored, they watched randomly designated "one-star" and "three-star" performers undergo painful punishment. The empathy centers of the subjects were noticeably less active when watching high achievers punished—apparently because we feel less empathy toward people whom we believe have bested us. To researchers, the subjects were jealous of the three-star performers, leading to them being less upset when these performers were tortured.

SELFISHNESS

"Greed is good," as Gordon Gekko said in the movie *Wall Street*. Certainly it's easy to surmise why selfishness could have survived as a social trait; in a world of limited resources, being nice may not be as important as being well fed. But we'd like to think it isn't so, that the societal rewards for working together and helping one another outweigh the value of any individuals grabbing all the good stuff. We hope that if selfishness is natural, civilization is the antidote.

According to recent research, altruism does indeed seem to be our default state, with selfishness the result of specific brain regions inhibiting our better nature. In one recent Hungarian study involving a cooperation game, when people who'd previously tested positive for Machiavellian traits encountered someone playing by the rules, their brains got excited. Specifically, they showed unusually high activity in the **dorsolateral prefrontal cortex**, which handles inhibitions, and the middle temporal gyrus, which handles creativity. In this case, creativity plus lack of inhibitions equals deviousness. It was as if they smelled opportunity and were actively trying to find ways to exploit the situation.

HATE

It's perhaps the most destructive of all our feelings—so what's its value? Why hasn't it been cleansed from our gene pool? Because hate is good: That's according to psychiatrist Anna Fels, who explains in *The New York Times* that we have valuable "hate circuitry" in our brains, which partly overlaps with our "love circuitry." This neuronal configuration is theorized to have evolved to preserve the cohesion of early social groups and emerging societies. To hate outsiders is to bond internally. It's a self-preservation mechanism for groups.

The Foundation of Belief

What does it mean to believe in something—in the prospects of an investment, in a divinity, in the faithfulness of a partner? This high-level abstract thinking is a recent development, from an evolutionary standpoint. At its simplest, belief can be defined as believing, without proof, that something exists. It's an opinion, not a fact, but it's an important opinion that helps define you. Maybe it's precisely because a belief is unprovable that it seems to invite emotional investment, as can be witnessed in any comments section online. People in general are much more passionate and excitable when discussing what they believe vs. what they know and can prove.

Think about the social issue you care about the most. Now imagine you're defending this most cherished topic around the Thanksgiving table. Then, your dad refuses to pass the turkey; he couldn't disagree more. Because the disagreement can't be proven away, tempers flare. You've dug in. Both of your brains are now firing on all cylinders. Your amygdalae, and his, are revving up with emotion; you're bracing for conflict. He's already there. Your insular cortex steps in: Heeding your amygdalae's warnings, it labels your dad's stance a threat to your very sense of self. This is going nowhere. Who's up for another drink with dessert?

A 2007 UCLA study discovered something intriguing: Our brains treat opinions—the things we know but can't prove—as if they were facts. Specifically, we use the same neural processes to consider our beliefs and opinions as we do to evaluate fact-based and mathematical statements. As study co-author Sam Harris explains, researchers found that "while evaluating mathematical, ethical, or factual statements requires very different kinds of processing, accepting or rejecting these statements seems to rely upon a more primitive process that may be content-neutral." Our opinions and ethical stances may be constructed out of thinner stuff than objective facts, but to our brain, they're "facts" as real as the capital of Oklahoma or the square roots of nine. Accordingly, our beliefs become integrated into our personal identity, so we feel threatened when they are challenged and dig in as if we were being physically threatened.

Disgust determines how close we get to people. It determines who we're going to kiss, who we're going to mate with, who we're going to sit next to.

–*Disgustologist Valerie Curtis*

"E pur, si mueve."

("And yet, she moves.")

Dr. Michael Shermer, in his book *The Believing Brain*, argues that humans are hardwired to search for belief systems: We evolved the capacity, and the desire, to connect the dots of the world into meaningful patterns. It's how we've always explained the world to ourselves—arguably the start of both religion and science—and in continuing to build out our increasingly detailed construction of the world and what it means, we consciously and subconsciously try to extend our chosen patterns and seek out confirming evidence to support what we've chosen to believe.

How Religion Changes Your Mind

Religion and science have sparred for centuries; they haven't always been called "religion" or "science," but the two concepts have been at war for all of human existence. If science is an effort to explain the world through facts and reason, religion, for many, offers an alternative worldview that's inclusive of unprovable beliefs.

Many used to assume there was a "**God spot**," some region in our brains specifically dedicated to faith, belief, and religious thought. But the notion was itself, ironically, faith-based, and recent studies have debunked the idea, demonstrating that the thought processes we think of as specifically religious in nature in fact simply hijack parts of the brains we were already using for other things. For example, according to Jordan Grafman, co-author of a relevant National Institute of Health (NIH) study cited in *Scientific American*, remembering a religious experience activates the same brain regions as remembering a recent meal.

Our concept of God is similarly treated in a rather mundane way by our brains. Humanity's new ability to monitor regional brain activity patterns while we're thinking about specific things reveals that thoughts of God aren't processed in the places where we have abstractions like love and peace, like one might think. Rather, according to fMRIs taken of people during prayer, our brains categorize God the way they categorize other people. Whatever your belief system, praying lights up your brain in the same way it does when you imagine meeting a celebrity or having a conversation with a historical figure. Nothing in the brain developed to communicate directly with abstract beings on another plane of existence, so the brain has to make do.

Proceeding from the general idea that repeated experiences can reinforce and strengthen particular neural pathways, the practice of religion changes your brain in a number of ways. "The more you focus on something—whether that's math or auto racing or football or God—the more that becomes your reality, the more it becomes written into the neural connections of your brain," said University of Pennsylvania neuroscientist Andrew Newberg in an interview with NPR. When you're compassionate and helpful and love thy neighbor, when you pray or participate in religious ceremonies, when you listen to sermons and try to apply them to your life, all these behaviors structure your plastic brain around them—they literally start to make up your mind.

The good deeds associated with religion, for example, can mold the mind through pleasure. They activate the same pleasure center, the nucleus accumbens that rewards you with dopamine during sex or a chocolate binge. It isn't just that giving feels good. It's that the expectation of feeling good drives later behavior—so much so that researchers have actually demonstrated that people's generosity is a function of the amount of chemical reward their brains dole out. Looking at the degree of activation in the nucleus accumbens, "you can actually measure how much activation there is and predict with some degree of accuracy how much they're going to give," says Dr. Bill Harbaugh, coauthor of one such study at the University of Oregon.

Is Faith Good For You?

There's a great and growing body of evidence that it is. Generally speaking, positive outlooks and meditative contemplation are both associ-

Quoth Galileo Galilei, in history's most defiant mumble, after being forced to renounce (before the Inquisition, in Rome) his heretical assertion that Copernicus was right and the Earth revolved around the sun. Just 350 short years later, Pope John Paul II admitted the church had been wrong.

Why can't we agree on the facts? Psychologists suggest that rather than unifying us, facts often drive an even deeper partisan divide. Humans have a **confirmation bias** that interprets new information in our favor, even if it challenges what we already believe. What can fix this? Scientific curiosity: People with this trait are more likely to seek out and genuinely evaluate contradictory perspectives.

ated with better health, and more particularly, studies have shown that people who profess a religious faith get an immune system boost and lowered blood pressure, and respond better to treatment for depression. Collectives of the faithful, like churches and synagogues and mosques, have long provided tangible benefits to society, from policing civilization-strengthening morality to promoting specific works like community building and grief counseling for families. But faith has clear benefits for the individual as well, providing for many a sense of inner strength and outward belonging.

There are caveats, though. For one thing, the beneficial effects of religion seem to be entirely dependent upon the nature of your relationship with God. In one study cited in *The New York Times*, subjects who viewed God as loving and available experienced calming effects and better health when they prayed; but for those who viewed God as remote, praying only added psychiatric stress. Chronic low-lev-

el stress is a big factor in the atrophy of the hippocampus, where new memories are formed, and researchers surmised that in this case it's the fear of God causing the stress. A 2011 Duke study found that people whose religion was defined by a "born again" or conversion experience had more stress and hippocampus atrophy than those who didn't identify as religious or whose religion had been stable.

Are Atheists Smarter Than Believers?

It's a seriously inconvenient truth, but the answer appears to be yes. In a meta-analysis of more than five dozen studies of intelligence and belief spanning nearly 100 years, researchers found that atheism correlated with higher IQ in more than 80% of the studies. In a separate Pew Research study, scientists were twice as likely as the population at large to say they didn't believe in God or another higher power.

Obviously, a great number of very intelligent people have also been very religious, including Charles Darwin and countless others. But on the average, intelligence seems to correlate negatively with religiosity. The question is why? Some theories include:

1. A worldview built on reason alone may provide some of the same fulfillment, self-regulation, and inspiration that religion provides for others.

2. Highly intelligent people tend to be non-

DOES **MEDITATION WORK?**

The question's been around for generations: Does meditation produce measurable benefits, or are practitioners fooling themselves? It's not as hippy-dippy as you might have been led to believe. Scientists looking at meditators (particularly, practitioners of MBSR, or Mindfulness-Based Stress Reduction) have found it can reduce mind wandering and help you snap back more quickly to the topic at hand. Overwhelmingly, the evidence says the practice can, at least in some cases, rebuild the brain in valuable ways. In one interesting study in mindful meditation conducted by Massachusetts General Hospital in 2011, subjects who quieted their mind and focused on breathing—without praying to a specific god—showed an increase in hippocampus density, which correlated with improved

ability to create new memories eight weeks later. More recently, a study examined subjects taking an eight-week meditation course focusing on two distinct types of mediation (while a control group simply experienced a general health seminar). After the course, the groups that meditated showed a greater density in gray matter around the hippocampus, and had less activity in the amygdala when shown negative images—a sign to researchers that they were better able to cope with stress. Long-term meditators claim to be better able to regulate emotions and control activities like quitting smoking, and meditation has been shown to reduce symptoms of pain, anxiety, and depression by roughly the same amount as antidepressant medication.

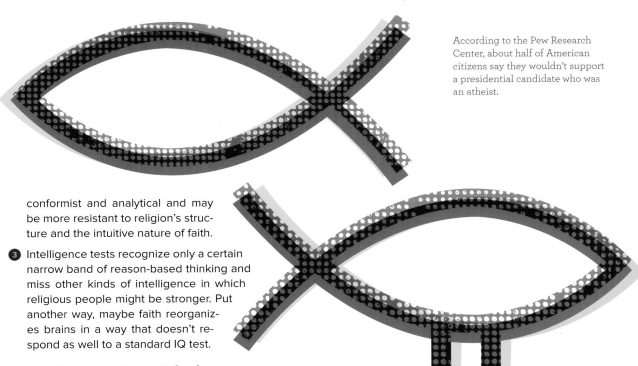

According to the Pew Research Center, about half of American citizens say they wouldn't support a presidential candidate who was an atheist.

conformist and analytical and may be more resistant to religion's structure and the intuitive nature of faith.

3 Intelligence tests recognize only a certain narrow band of reason-based thinking and miss other kinds of intelligence in which religious people might be stronger. Put another way, maybe faith reorganizes brains in a way that doesn't respond as well to a standard IQ test.

Cooperation: How Group Behavior Emerges

Cooperation is the great multiplying factor for intelligence, human or otherwise; animals can achieve great and terrible things when they unite behind a single goal. Think beehives and armies of ants, wild dogs hunting in packs, thundering herds of buffalo blackening the plains, a flock of seagulls. (No, not that guy with the hair...the other kind.) And for humans, with our superior intellects, the results, and the intense feelings gleaned from working together, can be much more profound.

Whether it's a multifaceted trained army acting in concert or corporate boardrooms and think tanks collaborating to create and build new ideas, we've long recognized that working together yields better results than the sum of individual efforts. Even in cases where society at large might take issue with the collaborative effort—like with the terrorist group ISIS or a hacker collective working to loot online banks—there's no denying that human cooperation yields outsize benefits.

Cooperation engages the animal brain more fully than solo activity. A 2011 Johns Hopkins University study looked at the song patterns of plain-tailed wrens, where male/female pairs often sing cooperatively in an A-B-C-D pattern with the male singing the A-C parts and the female the B-D parts. The birds' brains show increased activity throughout the duet (not just during their own singing) relative to neural activity when singing alone. Our brains are more active and engaged when we're cooperating. Even 3-year-olds, in another study, preferred to cooperate 78% of the time when given the choice. We get chemical rewards for this behavior: Our hypothalamus fires off oxytocin, which provides a feeling of warm comfort.

But human cooperation doesn't always lead to greatness. Bad ideas that an individual might reject out of hand—riots after sports championships, conspiracy theories that won't die, institutional racism—can spread quickly and persist when reinforced by the presence of other humans to share the responsibility. Dr. Shermer's assertion that the human brain is wired to find patterns in the world and ascribe universal meaning, even if it has to ignore facts to do so, may make us susceptible to sloppy thinking and the abandonment of reason and responsibility. Stress and trauma can be hard to process, and things like terror attacks or even job loss can send the brain scrambling to find reasons, causes and effects, and someone to blame. When a deserving target can't be found, we sometimes substitute. It's nothing to be proud of.

According to some scientists, some group behaviors we take issue with in the modern world may have had more rational explanations in our not so distant past. The xenophobia, or fear of strangers, that we find so distasteful today may have been much more helpful historically, when social groups regularly had to protect themselves and their scarce resources against the incursions of foreigners, says University of Minnesota primatologist Michael Wilson. This made sense in our past, when those outside "our" group probably wanted to do us harm. Today, we know that people of a different color or political mindset are unlikely to hit us over the head and take our food. But biases backed up with chemical rewards can be hard to overcome by force of will.

What Happens During a "Religious Experience"?

Intense religious experiences have been compared to certain types of epileptic seizures in terms of their effect on the brain. In his 2010 book *The Spiritual Doorway in the Brain: A Neurologist's Search for the God Experience*, Dr. Kevin Nelson attempted to ascertain the physiological cause behind the "tunnel of light" commonly reported during near-death experiences. In his view, the "tunnel of light"— the idea that people typically see a pathway to heaven right before they die—is really just our brain trying to make sense of the last gasp pulses sent through our visual cortex thanks to compromised blood pressure around the eyes. We see something unprecedented, and, as usual, the cortex tries to come up with a sensible explanation.

The widely reported phenomenon of your "life flashing before your eyes" can be explained similarly by a sort of relativity effect. Adrenaline is released during traumatic experiences, which speeds up mental processing—you may experience this as time seeming to slow down. Your brain may be rifling through past memories, trying to find an analogous situation to help you save yourself—and because it's doing so at a speed your consciousness doesn't typically reach in your everyday life, it feels like an otherworldly experience.

When you describe that fresh-baked cronut dripping with cinnamon sweat as a "religious experience," this colloquial usage may, in fact, be close to its actual impact on our brains. As the late neurologist Oliver Sacks described in *The Atlantic*, when people convince their own brains that they're hearing angelic voices or seeing faces in the light, they actually cause their sensory systems to deploy. "These yearned-for voices and visions have the reality of perception, and this is because they activate the perceptual systems of the brain, as all hallucinations do." You think you're hearing angels, but in reality you're hoping to hear angels—and your sensory systems try to comply.

Why Liberals and Conservatives Can't Talk

In America they're Republicans and Democrats; in 19th century England they were Whigs and

CAN PRAYER **REALLY HEAL?**

Lord only knows.

Those who claim to have been healed by a minister's touch may not be wrong; whether or not there was divine intervention may be another matter. What's likely at play is a psychosomatic effect not unlike when patients are given a placebo pill, which can in some cases help negative symptoms lessen or disappear even though it's only the patient's mind doing the healing. There's a reason that multiple sclerosis (MS) is often called a "quack magnet"— its symptoms and impairments can sometimes come and go without explanation, leading some to feel they've been "healed" by spiritualists or "miracle worker doctors."

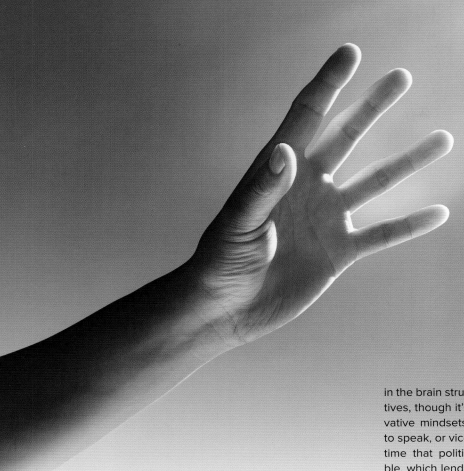

in the brain structures of liberals and conservatives, though it's not yet clear whether conservative mindsets build conservative brains, so to speak, or vice versa. We've known for some time that political leanings are partly heritable, which lends credence to the idea that the brain structure you're born with at least partly determines your ideology. Being "born liberal" or "born conservative," if the difference is significant, might help explain why it's so difficult to change someone's mind along liberal and conservative lines.

Are there structural differences? To find out, psychologists started by codifying the ethics that underlie political positioning. NYU professor of social psychology and morality expert Jonathan Haidt established a baseline of six basic pillars of moral thought, common across all religions and cultures, and tested self-identified conservatives and liberals to see where they stood on each. His pillars are: Care/Harm, Fairness/Cheating, Loyalty/Betrayal, Authority/Subversion, Sanctity/Degradation,

Tories. But across cultures and throughout history, there have been two main opposing forces in human political society: liberals and conservatives. Liberals want to embrace innovation, include everyone, and help the fallen and disadvantaged along. Conservatives want to move cautiously, reward hard work, champion individualism, preserve group structures, and defend against invasion and disruption. Though the names are most often used to reflect the organized groups that espouse these belief systems, liberalism and conservatism are really ideologies, either of which can be dominant in an individual or a larger group.

Interestingly, there are meaningful differences

Religious Americans are more likely to engage in unprotected sex and have higher rates of divorce than atheists and agnostics, especially in the "Bible Belt."

WHY CAN'T WE BE **FRIENDS?**

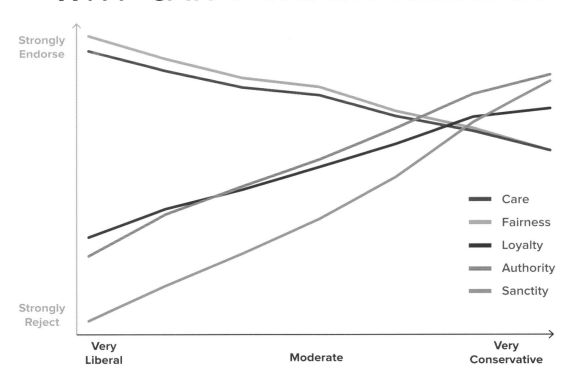

Strongly Endorse

Strongly Reject

Very Liberal   **Moderate**   **Very Conservative**

Legend:
- Care
- Fairness
- Loyalty
- Authority
- Sanctity

Psychology professor Jonathan Haidt of NYU, found that while conservatives uniformly valued each quality, liberals placed a much higher value on "Fairness" and "Care" relative to the other tested pillars, and a lower value on "Sanctity."

and Liberty/Oppression (which was omitted from the questionnaire whose data produced the graph above).

Haidt found that the differences between the two viewpoints really boil down to a differential weighting of basic ethical ideas. While conservatives assigned all the pillars roughly the same values, liberals placed a much higher value on "Fairness" and "Care" relative to the other tested pillars, and a lower value on "Sanctity." (Think of marriage policy, where fairness-minded liberals want to open marriage to the LGBTQ community and reject conservatives' "sanctity of marriage" argument out of hand.)

These differences are also evident in the brain's anatomy. Liberals tend to have a larger and more active anterior cingulate cortex, useful in judging conflict and error. Conservatives tend to have larger-than-average amygdalae, involved in emotional memories. In the face of an ambiguous situation, conservatives may process information with a stronger, more "visceral" emotional response and are far more likely to prefer safety and feel threatened by change. Liberals, on the other hand, are more

comfortable with change and may seek it out even at the expense of comforting stability.

Our neural network structures diverge so starkly along party lines that we can, in fact, actually predict a person's political leanings, with 95 percent accuracy or more, from a single fMRI image. Computational psychiatrist Read Montague showed conservatives and liberals four kinds of images: pleasant, neutral, threatening, and disgusting. The disgusting shots (including gruesome animal carcasses) were the most telling, provoking such a visceral reaction in conservatives that their affiliation could be predicted with near certainty by looking only at fMRI images of their reactions to these shots.

In a 2008 TED Talk called "The Moral Roots of Liberals and Conservatives," Haidt claimed you can even predict political ideals based on people's reactions to seemingly "neutral" questions about art or travel: Liberals, more cortical and analytical in their thinking, are on

One study found that devout Mormons have heightened activity in their nucleus accumbens, the same reward system that's activated during the enjoyment of love, music, and drugs.

THINK ABOUT THIS

average more open to new ideas and new experiences; conservatives prefer experiences that are familiar, as reflected in their life choices. This means, sadly, that some of the antagonism between the sides of the aisle may be structurally unresolvable. Morality "binds people together into teams that seek victory, not truth," says Haidt. "It closes hearts and minds to opponents even as it makes cooperation and decency possible within groups."

How Do People End Up in Cults?

Religion can help people feel fulfilled, gain self-control, and build family and community focus. Unfortunately, that reward cycle can be subverted by greedy and diabolical players who know how to manipulate and exploit. The world's a mean and complicated and, frankly, exhausting place, and there's something undeniably tempting about the prospect of handing over responsibility for some—or all—decision-making to a confident authority figure. And sometimes one shows up, right when you're most vulnerable—a self-styled visionary well-versed in how to tap into your self-doubt and susceptibility; one who understands your desire to be free of anxiety and care, who knows how to give you a reason to believe and go on, and who gives you the context of a community that welcomes you.

Congratulations: You just joined a cult.

What's happening on a neurological level? It's difficult to apply the scientific method to studying cults, since recreating brainwashing conditions for research purposes might be ever so slightly unethical. But some of what researchers have discovered about cult behavior suggests it's less about religion and faith and more in line with our basic animal needs as humans. John G. Clark, Jr., an assistant clinical professor of psychiatry at Harvard University Medical School, found back in the early '80s that while there isn't one specific profile for your typical

Psychiatrist Gail Saltz says that conservatives tend to have larger amygdalae, the brain's fight-or-flight center, where emotional information is processed.

cult member, a lot of the cases he studied were kids who were intelligent, came from sheltered environments, and had some religious upbringing or experience but who were not devout. What cults offered wasn't belief but friendship, identity, security, and respect. The people who are targeted most frequently tend to be in the midst of some transitional moment in their lives: loss of a loved one, career failure, divorce. Something needs to replace what was taken or lost, and a quick-thinking cult leader can step into the void.

The first order of business for a success-minded cult leader, often, is to cut followers off from the outside world. The cultists become emotionally insecure and even more reliant on the leader to provide for them. They become "brainwashed."

What drives people to form or lead cults, beyond the obvious narcissism? According to a former Moonie—a follower of legendary cult leader Sun Myung Moon—named Diane Benscoter, cult leaders become addicted to the money and power their flocks provide, and their brains react to it the same way they would

to drugs or alcohol addiction.

In a 2009 TED talk, Benscoter used the term "viral memetic infection" to describe what cults do to your brain: In the Us vs. Them mentality common inside cults, self-destructive ideas can pass from person to person and become so powerful members can feel compelled to follow through with them. Benscoter says people are most vulnerable during periods of crisis or doubt, just as someone with a vulnerable immune system is more susceptible to disease. In a similar vein, Oxford psychologist Kathleen Taylor warns not to think of brainwashing as some "magical weird process." Rather, she asserts, it's a mundane application of psychology and neuroscience—coercion, essentially, taken

Why do sports fans go crazy after big games? Psychologists suggest it may have much to do with "BIRGing" ("Basking in Reflected Glory"—your team becomes an extension of your own identity), mass disinhibition, and exhibitionism that draws in even the most introverted fans, whether at the stadium or at the local sports bar. Add the accompanying surge of testosterone and influence of alcohol and you'll see victory (or defeat) riots.

THINK ABOUT THIS

to the extreme. Repetitive tasks, whether meditation or forced sleep deprivation, change our brains over time.

A phenomenon often associated with brainwashing is **Stockholm syndrome** (see sidebar), but the condition has a lesser-known, but no less fascinating, alter ego: **Lima syndrome**, in which captors come to sympathize with their captives. The name is born from a Peruvian psychiatrist, one of hundreds held hostage in a 1996 takeover of the Japanese Embassy in the capital, Lima. Dr. Mariano Querol was able to successfully appeal to the captors' sense of compassion through counseling and morning aerobics. The doctor spent the nearly three weeks in captivity reading to them, including their favorite: Gabriel Garcia Marquez' *News of a Kidnapping*. That was enough for the lead captor, a businessman deeply in debt, to spring for a free cab ride home for the doctor—but only after collecting his ransom money.

Consider now: Did Belle develop Stockholm syndrome or did the Beast develop Lima syndrome?

Herd Mentality

Cults are one example of a phenomenon where groups of people subjugate themselves to the will of another. But a related effect is when people abdicate their choices not to one powerful individual, but to nobody at all—to the intuited will of a leaderless group. This is mob behavior, and it can lead to some truly fascinating and dangerous outcomes, ranging from riots to stock market collapses.

Let's start with the concept of herd mentality and consider that you may, in fact, be a born follower, thanks to your brain's insatiable craving to explain away uncertainty. Researchers at the University of Leeds, led by Professor Jens Krause, used the "snake test" to illustrate this. A large group of students were asked to walk randomly around in a hallway, and not to speak or gesture to one another. However, unbeknownst to the group as a whole, a few select students had been secretly instructed to walk in a specific pattern. Before long, a "snake" of students began to fall in line behind the few students who'd been told to walk in the pattern. "In most cases," reported Krause, "the participants didn't realize they were being led by others." Researchers began increasing the size of the group to find out just how few students with special instructions could control the oblivious masses. They discovered that as few as five

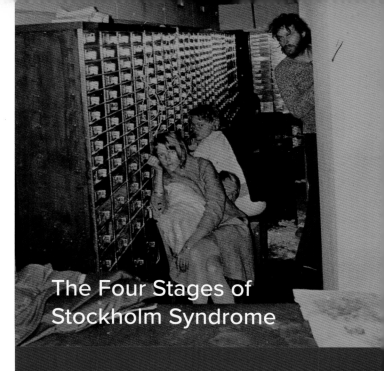

The Four Stages of Stockholm Syndrome

THE BYSTANDER EFFECT **MEETS THE ZOMBIE STRATEGY**

We are less likely to act if we are in a crowd. In one-on-one crisis scenarios, a bystander will typically recognize her responsibility and pitch in, but the more people that gather around her, the more that responsibility is diffused, and the more likely she will do nothing and wait for someone else to step up. This situation was made famous in the Kitty Genovese case, where Ms. Genovese was stabbed to death outside her apartment and many witnesses heard her cries but did little or nothing, perhaps hoping someone else would try to stop it. (Some of the details were later disputed.) Writer Melanie Tannenbaum wrote in *Scientific American* about using deindividuation as a strategy during a "zombie marathon," a recreational event where people acting as zombies attempted to take flags from the belts of "human" runners. Tannenbaum found that running around individual zombies didn't work, but that running straight into a pack of them was a better strategy, because each zombie thought somebody would grab your flag...so, often, none of them did.

directed students could lead 95 on a subconscious wild goose chase.

Herd mentality has led to some truly bizarre historical incidents. The dot-com boom of the 1990s was a textbook example, where newly minted Internet ventures with no discernible business models were able to prey upon the uninformed public's insatiable desire to climb aboard the Next Big Thing, a model that was as irresistible as it was unsustainable. Such speculation-crash cycles have been going on for centuries, including a bizarre tulip craze in Holland in the 1600s that had individual tulip bulbs selling for as high as $400,000 apiece in today's dollars, until the whole thing crashed.

Why are we susceptible to this urge to abandon reason and join the crowd? In part, it's simply easier; in part, it's exciting. Psychologist Tamara Avant calls this "deindividuation"—we lose our "self" in the crowd, which frees us from our inhibitions and responsibilities and liberates us momentarily to "go with our gut" and act outside the normal rules of society. This can lead to much worse behavior than bad investments. Activities like lynch mobs and riots—which may begin as peaceful protests or innocent celebrations, but can quickly turn dark when a few break ranks and start wrecking and looting—can spread rapidly once it seems like everyone's doing it. They can be truly terrifying, even for participants, who can walk away wondering why they did the things they did.

In modern times, this "deindividuation" reaches a natural apogee on the Internet, where, thanks to the further democratizing power of social media, we simultaneously have direct phone-level access to the entire world—and can be as anonymous as we like. If you've ever wondered what shedding inhibitions and responsibilities looks like, visit the comments section of any controversial topic—say, the 2016 U.S. presidential election—people who are doubtlessly polite and reasonable IRL (in real life) become the most caustic, venom-spewing trolls once they can hide behind avatars. The price of the freedom of anonymity is a constant, disturbing look at humanity's very worst impulses.

But though the digitally connected world may sometimes reveal our darker impulses and appetites, the future isn't bleak—not by a long shot. In the next and final chapter, we'll take a look at some of the truly mind-blowing advances in neuroscience, robotics, and other disciplines that will shape our lives and minds in the years to come. ✧

Advertising Techniques vs. Cult Tactics

How can cult followers be coerced into taking part in something so profoundly against their own self-interest? Pretty much the same way advertisers get you to buy products you don't need, according to Oxford University's Kathleen Taylor. Consider the techniques:

Repeating words and phrases: Advertisers and cult leaders alike reinforce their messages—literally strengthening the memory paths of their propositions in your head—through patient, mantra-like repetition of key phrases, like "I am the path" or "Snickers really satisfies."

Tribalism: We all have a need to belong, a primitive response that harkens back to when communities needed to work together to survive. Now that we live more comfortably, we seek that belonging and protection by declaring ourselves "Team this" or "Team that"—and it's effective, whether you're an Apple vs. PC person, Yankees vs. Red Sox, Coke vs. Pepsi, or fringe cult vs. unenlightened outsiders.

Doubt: Planting the seed of doubt in a person's mind about his past choices is a key to both marketing and brainwashing. Cult leaders might point out the persistence of his unhappiness. And how many ads are built around the same basic premise: "You've tried the rest—now try the best."

CHAPTER 9

SMART EVERYTHING, THE AI AGE, AND WHAT'S NEXT

Of all the chapters in this book, this is the one that should fill you with the most inspiration— and dread.

Man and machine have been inching together arguably since the dawn of tools, but the alarming pace at which our tools have begun to surpass us in capabilities long thought to be unique to humans is at once truly exciting and deeply terrifying. Artificial intelligence (AI) today is estimated to be at about the intelligence level of a 4-year-old child, and it's improving much faster than that child, with Deep Blue (born 1993), Siri and Watson (born 2010), Cortana (born 2013), Alexa (born 2014), Google Assistant (born 2016), and no doubt many more under wraps, all in furious continuous development while you blithely binge-watch Netflix.

In fact, luminaries like Stephen Hawking and Elon Musk are already warning about **The Singularity**, a point in time when artificial intelligence will outstrip the capabilities of human intelligence and start to take its evolution out of our puny meat-hands. The Singularity may no longer be theoretical: Google's head of engineering and longtime futurist Ray Kurtzweil predicts we will cross this horizon in the year 2029.

In other words, you may still be alive when our machines become smarter than us. How do you feel about the End of the Human Era, when we're no longer—never again, in fact—the smartest kids in class?

MIND BENDERS

1997
Year in which a computer first beat the best human at chess

2011
Year in which a computer first beat the best human at Jeopardy

2029
Year when AI's been predicted to surpass the intelligence level of an adult human

1.5 billion articles per year are now written by software, not people, and the number is growing.

But artificial intelligence is just one of many technological advances impacting the future of your mind. Your home, car, and other objects in your life are already "smart" and digitally connected. They accept your voice commands and have become extensions of your being. We're right around the corner from manipulating our memories at will, dramatically enhancing our ability to learn, curing bad habits while sleeping, and all sorts of other "magic." **Neuromorphic engineers** are working to simulate a complete human brain inside a computer, while bioengineers are busy creating original brains from stem cells. Other promising experiments are raising the possibility of uploading your own brain, in its entirety, into digital storage—otherwise known as *immortality*, perhaps—creating a host of new ethical challenges beyond anything we've ever dreamed of.

Buckle up!

Your Head vs. Your Heart

Jeni Stepien was about to get married in 2016 and keenly missed her father, who'd passed away 10 years earlier. It didn't seem quite right getting married without him. But Dad had been an organ donor, and so Jeni tracked down Arthur Thomas, the man who'd received her father's heart, and asked *him* to walk her down the aisle so she could listen to his heartbeat—her father's heartbeat, that is—on her special day. Purely symbolic? Maybe, but the video of the event went viral and spread to news outlets all over the world. Clearly, the sentiment touched a very human nerve.

Intellectually, we know the heart is just a muscle pumping away insensibly; most of us know a father cannot meaningfully be replaced by his organ recipient. But that's just our heads talking. For centuries humanity has found it easy enough to believe the heart to be the seat of romance (possibly because of the "heart fluttering" brought on by romantic nervousness),

and the evidence is still there in our idioms: "matters of the heart," "follow your heart," and so on. You place the ring on your beloved's ring finger because of an ancient belief that the ring finger housed a unique artery running, in the words of Bryan Adams, "straight from the heart." (It isn't true; all fingers share a similar circulatory structure.) We know better, now, but what you know isn't always what you feel.

And that's where it gets interesting.

Because when you get a heart transplant, *you're still you*—organ recipient Arthur Thomas in no way became the bride's father. But what about a head or brain transplant? If you and I were to decapitate one another in a disastrous swordfight, and ER doctors were only able to save your head and my body, the resulting

90%: Accuracy rate of IBM Watson's lung cancer diagnoses, on par with the accuracy of human physicians.

Franken-hybrid would be mostly me, by mass; if it had children, they would share my genes, not yours. But I think we'd all agree that this creature, with your head and my body, is YOU, not me. It wouldn't really be a head transplant at all—it would be a body transplant. The identity goes with the head.

As you might have guessed, this isn't just a theoretical question: The first human head replacement surgery is already in the works—details below. Just one of the 1,001 insane breakthroughs, and agonizing ethical dilemmas, coming our way. Here are six more.

Breakthrough: Telepathic Communication Begins

Are you ready for direct brain-to-brain communication, where you can send your thoughts directly to your spouse, boss, or the Starbucks barista? The party trick of the coming decade took a big step forward in 2013, when University of Washington researchers Rajesh Rao and Andrea Stocco first successfully demonstrated human-to-human mind control. Rao played a basic shooter video game while an electrode-studded **electroencephalogram (EEG)** cap read his brain's electrical activity. Across campus, Stocco sat beneath a **transcranial magnetic stimulation (TMS)** coil positioned over his left motor cortex, the region responsible for hand movement. When Rao imagined firing the cannon with his right index finger, his brain activity triggered Stocco's right index finger to click his fire button. The involuntary impulse felt, to Stocco, like a nervous tic. This super-simplified thought transfer isn't going to replace spoken language anytime soon, but as it develops, the possibilities—for remote soldiers automatically relaying field information and receiving commands, for telepathic cheating on exams, for police officers questioning witnesses—are endless.

Breakthrough: Linked Minds Accomplish Miracles

The hive mind is a powerful metaphor, and scientists have recently begun experimenting to bring it to life. The idea is to link two or more brains together into a network—a Brainet—that shares thoughts, memories, and functions. Imagine, for example, if someone struggling to regain motor control after an accident or an illness could "link up" with a healthy person and relearn the skills more quickly. With multiple brains wired together, these Brainets might well be powerful "organic computers" beyond anything we can imagine. In one experiment, three rhesus monkeys connected in this way moved a virtual arm better than any individual monkey; in another, two human gamers thus connected performed better together in a spacecraft navigation simulation than individually. The promise is powerful, but there may be new risks as well: Could destructive thoughts or a psychotic episode in one subject, for example, spread across a multi-brain collective like a computer virus?

Breakthrough: Mind-Controlled Prosthetics Walk the Walk

Our very first attempts at the human/machine interface were prosthetic limbs, and the ultimate dream is to one day replace amputated or malformed limbs with fully haptic (i.e., sensory-stimulated, feeling) prosthetics controlled by the brain like natural limbs. Could a quadriplegic patient control an exoskeleton and walk again? It's already happening. In 2011, researchers behind a project called BrainGate2, led by neuroscientist John Donoghue of Brown University, watched as a paralyzed woman named Cathy Hutchinson successfully manipulated a robotic arm using only her mind. By 2015, scientists in Germany and South Korea had developed a full exoskeleton that could be controlled by a non-invasive skullcap dotted with electrodes. The exoskeleton is bulky, and the sea of flashing LED lights might induce epileptic seizures, but it's a powerful start. Other robotic limb projects that don't tap into the brain could include revolutionary prosthetic legs developed by Hugh Herr, the head of the Biomechatronics group at MIT's Media Lab and himself a double amputee. Herr's robotic limbs are driven by self-contained, lithium-battery-powered motors designed to mimic the ankle (a notoriously delicate and dauntingly complex part of human anatomy); they store energy with each step down, propelling the wearer forward with each step up, as muscles and tendons do. Soon, a devastating injury might only mean a temporary inconvenience rather than a lifetime of diminished activity.

Breakthrough: The Internet of Things Goes to Eleven

Everything from your home security system to your electric toothbrush is fast becoming wireless-enabled and communicative; your appliances may be gossiping about you as we speak. But this is only Phase One of the Internet of things, and we're starting to get a sense of the true, boundless potential of this connected world. Take transportation, for example: Imagine cities paved with smart concrete that can alert service crews to widening cracks or weakening bridges before they fail, and cars that can talk to one another (to reroute around traffic snarls) and to the road itself (to adjust the tread or spacing of your tires for oil slicks or road hazards). Humans will be in on the fun, too: A group called Blackrock Microsystems is developing a wearable neural interface that will let you, too, communicate with the Internet of things (e.g., so college students, at the crack of noon, can *think* the alarm clock off and the coffee maker on) as easily as the devices communicate with one another. Security concerns will multiply, of course, as all connected devices are potentially vulnerable to hacking. You know what they say: The way to a man's bitcoin is through his toaster.

Breakthrough: Researchers Start Printing the Brain Tissue They Need

The ability to 3-D print body parts and biodegradable plastic scaffolding that can temporar-

If movie nights at cat cafes are too passé for you, check out the Gallant Lab at UC Berkeley. Once you've settled into their fMRI, they'll put on a flick—and will use your brain activity to reproduce what you're seeing on a separate screen. (Keep it PG.)

THINK ABOUT THIS

From smart watches to Facetime to dog treadmills, Hanna-Barbera's *The Jetsons* proved to be more forward-thinking than its now-retro veneer may suggest.

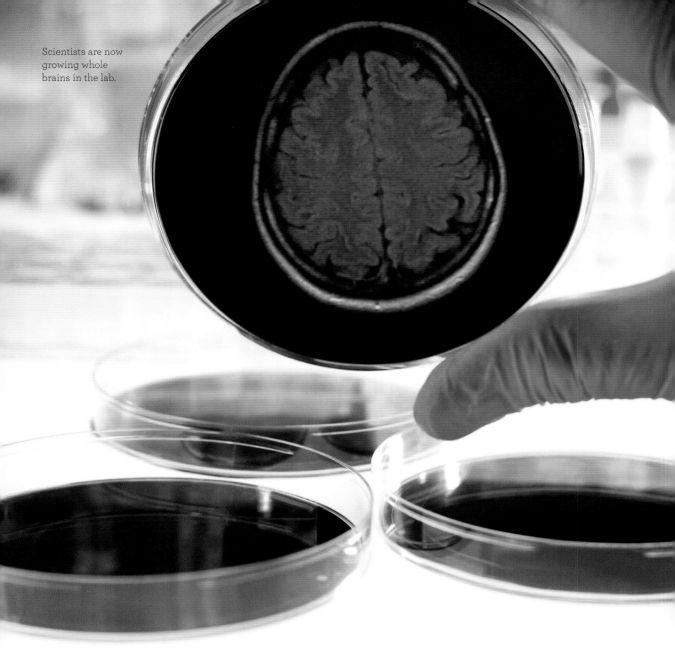

Scientists are now growing whole brains in the lab.

ily hold together new muscle and cartilage cells as they grow may sound fantastical, but it's already happening. That isn't the breakthrough. Researchers at the Wake Forest School of Medicine have developed the ITOP—which stands for Integrated Tissue and Organ Printing—system, which creates sponge-like temporary replacement parts specifically designed to break down and be replaced by the new healthy tissue as it grows. Brain tissue is a special challenge, but if we can actually build partial brains with a process like this, we could one day provide researchers with an endless supply of incomplete but reliable test brains that can be used to test drug interactions, track the progression of disease, and

explore other phenomena. Harvard University scientists have already used a form of this process to construct a gel mimic of a fetal brain and emulate the development of its cortical wrinkles and folds.

Breakthrough: Others Are Growing Whole Brains in the Lab

The trouble with researching the brain conditions that plague us, like Alzheimer's, strokes, etc., is that it's hard to tell what's going on in a brain unless it's living, and people are very attached to their factory-issue brains. But what if we could grow brains, separate from bodies, again purely for research? Mad scientists—er, researchers—at the University of Edinburgh found that they could grow **mini-brains** from stem cells in controlled conditions, and watch them form and grow. Since the bitty brains are made from the cells and DNA of donors, some with mental illnesses, scientists can track the illnesses' growth and development right before their eyes. A group from Ohio State University was able to grow a brain about the size of a pencil eraser, roughly equivalent to that of a 5-week-old fetus in the womb. The tiny brain is complete with 99 percent of the cells and genes that would exist in a human fetus, including the beginnings of a small spinal column and of an eye. Dr. Rene Anand, who led the project, dismissed any ethical questions by assuring the press that they didn't "have any sensory stimuli entering the brain. This brain is not thinking in any way."

Those are just a few of the major breakthroughs that are challenging our traditional relationship with our brain and mind. But advances in neuroscience and related fields are promising benefits in virtually every area of human endeavor, including some of the topics we've covered in previous chapters. Take a look at some of these interesting futures unfolding right now.

The Future of...Sleep

We know that some aspects of our busy modern life—like shift work and artificial lighting—are incompatible with getting the quality of sleep our bodies and minds need for good health. But does sleep have to be accomplished in the usual inefficient way, or do our sleep hours hold untapped potential for growth, healing, and enhancement? In the coming years, it may be that the idea of simply passing out for eight hours a night will seem as quaint and primitive as expecting leeches to cure tonsillitis. Researchers are investigating ways to make our sleep time more efficient, and our future may well be filled with controlled and productive dreams, malleable sleep schedules, and smart pajamas.

Some researchers are trying to find ways to simply reduce our natural sleep cycle, with efforts centered on a gene called DEC2, dubbed the "sleep gene." A mutation in DEC2 allows some people to get along fine with five to six hours of sleep, as opposed to the seven to eight hours usually required for optimal functioning. Scientists speculate that further modifying the gene could reduce or eliminate our need for sleep as we know it altogether—an idea that's reportedly piqued the interest of the military. We might instead be able to get by with napping: During short snoozes, the brain's right hemisphere remains extremely active, and researchers have discovered that a 20-minute nap can boost creativity and focus. Napping at work has already gained traction among several progressive businesses like Google, Zappos, and Uber, with some offering designated nap rooms or pods for staff members.

Another area under exploration is hibernation. NASA has long been experimenting with therapeutic hypothermia to help astronauts hibernate during long missions (as imagined in films like *Interstellar* and *Avatar*); a version of the technique is already in use in some hospitals to lower patient body temperature by a few degrees for a few days to induce a deep sleep conducive to epilepsy treatment. Separately, a study called Project Implicit, which began at Harvard University and was picked up by a

"I had dinner recently with a guy who bragged that he had only gotten four hours of sleep that night... I thought to myself 'If you had gotten five, this dinner would have been a lot more interesting.'"

– Arianna Huffington
Author, *The Sleep Revolution*

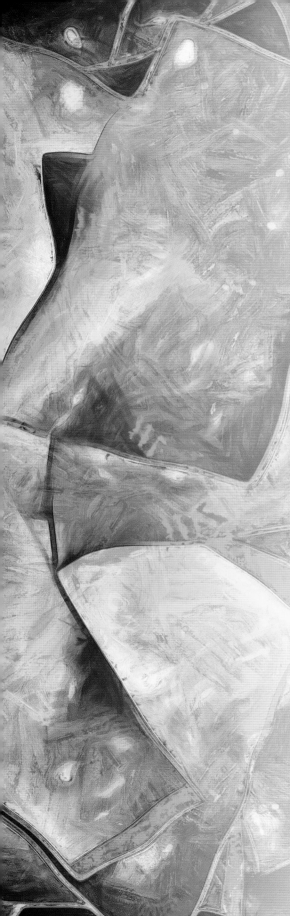

team from Northwestern University, offers hope that such periods of low-temperature hibernation might afford people an unprecedented opportunity to deprogram PTSD or unwanted racist or sexist thoughts. Using a mixture of sensory suggestions that combine pictures, words, and specific sound cues to, for example, change someone's negative associations with one gender or the other, researchers found that biases dropped as much as 56 percent in subjects who underwent the treatment during a hibernation session.

Speaking of accomplishing things while you sleep, let's talk about lucid dreaming. Currently, accessing the benefits of lucid dreaming means training yourself to be able to control your dreams. But what if you could accomplish the feat with a mild electric jolt instead? In 2014, Dr. Ursula Voss of the Johann Wolfgang Goethe University in Germany was looking at electrical activity during REM sleep and found a link between higher EEG activity and conscious awareness while dreaming. She discovered that giving sleeping patients a small jolt of electricity to their frontal lobe induced "self-reflective awareness" during dreaming—i.e., lucid dreaming. Because lucid dreaming allows some limited control of the dream environment, Voss' use of electrical kickstarts could potentially help people with PTSD or anxiety-induced night terrors actively confront their fears or even mentally change the outcome of a recurring traumatic event, to help their brains recontextualize the memory into something less crippling.

Finally, our approach to sleep will likely change with the rapid development of wearable tech. It won't be long before nearly everything wearable—from wristwatches to underwear—may be programmable and theoretically able to "speak" to us and actively assist us in living and performing well. When your pajamas join the party, they'll be able to work overtime while you sleep to gather data about your general state of health, documenting pulse, blood pressure, EEG patterns and more so that you can wake up to a complete medical diagnosis along with your morning coffee.

The Future of...Drugs

Ingesting the same drugs as everyone else and gyrating to a hologram revival of a long-dead rock star is *so* 2015. The experience revolution is coming, and it will be spectacular, fully immersive, and customized to your particular proclivities. Advances in brain mapping and gene sequencing could open the door for drugs tailored to the needs and desires of an individual user. And not only for recreation: Designer drugs could help with everything we do, from relaxing to learning to sleeping to forgetting. Later, as we get a better handle on the specific electrochemical brain activity these substances induce, the drugs may themselves give way to direct transcranial electromagnetic stimulation, allowing for drug-free, electrically induced highs. Imagine, in the hands of a particularly skilled DJ, a song paired with an electromagnetic pulse combination that lets an entire club experience an altered state without having to take any substances—an experience that can be centrally turned off, at the end of the performance, for a safe drive home.

For those who want *fewer* drugs in their lives—you rebels—there'll be solutions for that, too. To help children, addicts, and others who want or need to opt out, there'll be "drug vaccines" that prevent certain drugs—like, say, heroin—from interacting with your brain even if you ingest them. Similarly, alcohol could be reformulated to give you a buzz without the memory loss and hangovers, and other side effects could be reduced or eliminated as drugs can more carefully target only those receptors that make you relaxed and happy, not the ones that make you unsteady and woozy.

The Future of...Memory

Once in awhile, you can recall something you'd long ago forgotten—remembering a song from childhood, say, when you return to your old schoolyard. It seems no memories are ever truly lost, given the right trigger. And the Defense Advanced Research Projects Agency (DARPA), the folks who brought us the Internet, are now working with researchers at the University of Pennsylvania to focus on our brain's neural replay centers in the hopes of fully understanding our memories and achieving, ultimately, total recall. The multi-year program, called Restoring Active Memory, or RAM, will involve surgically inserting an implant, then jolting the brain with electricity to improve memory performance and recall. In late 2016, the team hit a critical mass of more than 1,000 hours of memory data drawn from more than 200 patients across the country. The implementation, however, is still projected to be years away, pending further study.

In a related study using transcranial magnetic stimulation (TMS), researchers from Northwestern University exposed subjects to weak electrical currents meant to stimulate the neurons in the cortex (and some, as controls, were merely *told* they were receiving such currents). Subjects who'd received the currents showed a 30 percent boost in the ability to recall information five days later. Results are preliminary, and we still don't know details like how long the effect persists, but the Northwestern

research team posits that transcranial magnetic stimulation increases communication between the hippocampus, the posterior parietal cortex, and the rest of the brain—raising the long-term potential to supercharge our ability to retrieve and relive old memories.

The Future of...Rock & Roll

Music, too, is undergoing its own technical revolution. Japanese singer Hatsune Miku, one of the world's fastest-emerging pop stars, doesn't write her own music, choreograph her own dance moves, throw tantrums, or argue about her contract...because she doesn't actually exist. She's a Vocaloid program, projected on-stage via hologram and surrounded by a live band playing "her" crowdsourced songs. Digital music is already much more fan-interactive than its analog predecessor; surprise albums are dropped overnight and half-finished works are released to an audience as raw material for remix and mashup possibilities. And Hatsune could be the Rihanna of this new world: a 100% malleable, knowingly artificial pop star, made for the fans by the fans and singing whatever they want to hear.

You may soon feel music differently, too: Wearable tech already helps the deaf feel the music, and its capabilities will soon expand to a wider audience: a vest that pulses to the music, a backpack that acts like an adjustable personal subwoofer, and more. Your experience, even at a concert with 50,000 people, will be unique to you.

The Future of...Sex

As we dig deep into understanding the neurology of sex, we're discovering that your electrochemical mind really drives the pleasure

Apparently Mark Zuckerberg's an *Iron Man* fan: He's developed his own home AI system, also named "Jarvis," voiced by Morgan Freeman. "He" can recognize visitors, control doors and A/C, start the toaster, and—most importantly—**shoot t-shirts out of a closet cannon.**

experience, much more than your naughty bits. And that means it's ripe for a virtual reinvention. Futurists predict that the explosive growth of virtual reality (VR) and haptic (feeling) technology will lead to a future where sex can be almost entirely mental, with just a few devices needed to simulate specific imagery, audio, and physical contact. Soon after Naughty America produced the first VR adult film, sex toys were given a high-tech boost. For the bold, there are now smartphone-paired his, hers, and couples' sex toys, where you and a long-distance partner can each feel what the other is doing down there; there's also underwear (Fundawear, from Durex) that simply lets you tickle your lover's nethers remotely, while they're sitting in that boring meeting. One smartphone app measures muscle response, body temperature, and heart rate to provide data on a person's ideal sexual experience; a group called Realbotix has created AI-driven robotic heads (and optional bodies) that would give haptic upgrades to the VR sex experience. Some experts, like sex and relationship therapist Dr. Laura Berman, see fully functional sex robots being the norm within a generation. Is the market ready? A U.K. poll from 2014 found that one in five residents would gladly have sex with a robot.

The Future of...Death

Spoiler alert: You're going to die. Or, at least, that's what we *used* to think. Humanity's been searching for the fountain of youth and the Holy Grail and other ways to live forever since the dawn of our history, and for the most part we've taken to believing there's no fix in this world: Dying is just a natural part of living. But what if it weren't so? There are curious suggestions growing that maybe it's not as necessary as we think. A sea creature colloquially called "The Immortal Jellyfish" can revert to a polyp state in times of stress, and reemerge in a genetically identical adult form later. Telomeres, the shoe-string-ends of your DNA sequences, get shorter every time your cells divide until they can't anymore, at which point the cells die. It's a serious roadblock for the immortality-minded, but adult lobsters produce a substance that repairs their telomeres, so they can live as long as 140 years, or at least until they're face-to-face with a pot of boiling water. Science has reinvigorated our ancient quest for immortality; as humans, we seem to be shifting from looking for supernatural help toward figuring it out for ourselves.

Here are some promising methods already on the drawing board.

METHOD #1: THE SYNTHETIC BIOLOGY SOLUTION

Could the key to immortality be hidden in our own bodies? Henrietta Lacks died of cervical cancer in 1951, but cells from her body were noted for reproducing quickly and for outliving normal cells, and were considered great candidates for experimentation. And so they have been: Lacks' cells have been kept alive for more than 60 years, forming an "immortal" cell line that's been used in thousands of medical experiments—they've been used for everything from developing Jonas Salk's original polio vaccine to researching AIDS. Today, scientists are experimenting with synthetic

biocircuits, grown from a person's own stem cells, that could be used to replace or repair damaged cells, or kill cells with mutations linked to cancer or aging. Similarly, DNA intervention could one day make our species more resistant to traditional threats like viruses and bacteria—researchers led by Dr. Farren Isaacs of Yale University and Dr. George Church of Harvard University, for example, have successfully rewritten the genetic codes of E. coli bacteria samples to be resistant to viruses and genetic mutation. Gene therapy could get very creative: Imagine if your body could start to glow like a firefly wherever cancerous cells were gathering, or if we could borrow some of the DNA-repairing abilities of the *Polypedilum vanderplanki*, a fly whose larvae can survive extremes of heat, cold, and dryness. Bodies that could continually repair and upgrade themselves might not help you live forever, but they could theoretically extend your lifespan a long time.

METHOD #2: THE BODY TRANSPLANT SOLUTION

Could your old head one day be attached to a healthy young donor body? It sounds like a scheme dreamed up by *The Simpsons'* diabolical Mr. Burns, but the operation is not only theoretically possible, it's already in the works. Animal head transplants (dogs and monkeys) have been carried out since the '70s, with the subjects often living for hours or days, though quadriplegic, as reconnecting the spinal cord and nerves has so far proved too daunting. But humans? And body control? Controversial Italian neuroscientist Sergio Canavero has said he believes it can be done by the end of 2017, and even has a willing subject: a young Russian programmer whose body is wasting away from a muscular disease. Other scientists are dubious of Canavero's timeframe, and the ethical implications couldn't be more complicated, redefining our sense of identity itself. Head transplant survivors would have two sets of DNA, for starters. Could we hold them responsible for crimes that *their* head or body had committed prior to the operation? If they carried children to term, would that make them natural parents or surrogates? And so on.

METHOD #3: THE BRAIN UPLOAD SOLUTION

If we get comfortable with the idea of consciousness being the core of our identity, and our bodies merely replaceable vehicles for our minds, the parameters of true immortality get a little bit broader. We can start to consider a **transhumanist** future, where our minds could live forever, escaping our mortal and flawed bodies by being uploaded into some sort of computer. The complication is fantastic—it would involve, at minimum, uploading the connectome, the trillions of connections among the neurons in our brains, not just the 86 billion neurons them-

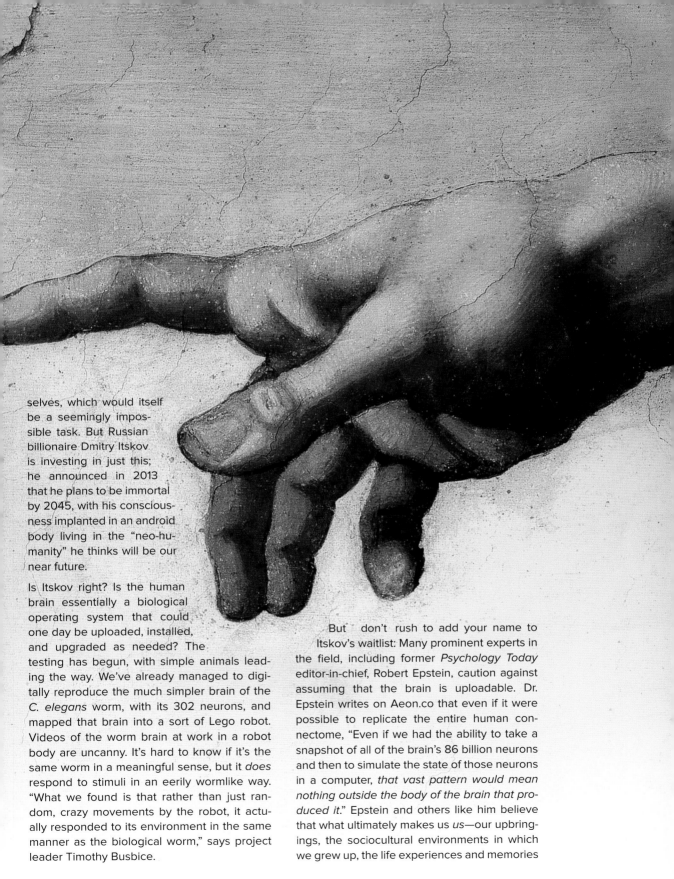

selves, which would itself be a seemingly impossible task. But Russian billionaire Dmitry Itskov is investing in just this; he announced in 2013 that he plans to be immortal by 2045, with his consciousness implanted in an android body living in the "neo-humanity" he thinks will be our near future.

Is Itskov right? Is the human brain essentially a biological operating system that could one day be uploaded, installed, and upgraded as needed? The testing has begun, with simple animals leading the way. We've already managed to digitally reproduce the much simpler brain of the *C. elegans* worm, with its 302 neurons, and mapped that brain into a sort of Lego robot. Videos of the worm brain at work in a robot body are uncanny. It's hard to know if it's the same worm in a meaningful sense, but it *does* respond to stimuli in an eerily wormlike way. "What we found is that rather than just random, crazy movements by the robot, it actually responded to its environment in the same manner as the biological worm," says project leader Timothy Busbice.

But don't rush to add your name to Itskov's waitlist: Many prominent experts in the field, including former *Psychology Today* editor-in-chief, Robert Epstein, caution against assuming that the brain is uploadable. Dr. Epstein writes on Aeon.co that even if it were possible to replicate the entire human connectome, "Even if we had the ability to take a snapshot of all of the brain's 86 billion neurons and then to simulate the state of those neurons in a computer, *that vast pattern would mean nothing outside the body of the brain that produced it*." Epstein and others like him believe that what ultimately makes us *us*—our upbringings, the sociocultural environments in which we grew up, the life experiences and memories

In 2016 ArsTechnica.com debuted "Sunspring," the first short film written entirely by an AI (one that named itself Benjamin).

that shaped us into who we are today—cannot be decoded, replicated, uploaded, or downloaded. This absolute uniqueness, he claims, relies on the interdependence of a particular body with a particular brain, and is simply not reproducible by any machine. But the jury is still out.

Artificial Intelligence and the Dangers of Playing God

In parallel with trying to extend or expand and upload a human brain, scientists have been working for decades on creating a computer analog of brainpower—artificial intelligence. If you use Siri to find the nearest pizza joint, or Netflix or Amazon's algorithm to choose your entertainment, or if you've ever had to "prove you're a human" to access a website, you're already interfacing with artificial intelligence. And it's starting to explode.

We've been talking about AI for years, of course—the problem proved much more intractable than we originally thought, which was secretly a relief to non-programmer humans everywhere. Early gains, like IBM computer Deep Blue finally besting humanity's finest chess grandmasters, were mostly computational, taking advantage of the computer's superior ability to look at thousands or millions of possibilities in microseconds before acting. With big data, machines can work a lot of what seems indistinguishable from magic, like predicting revolutions based on deteriorating local Twitter sentiment, or flu outbreaks based on facial tissue purchases.

Other aspects of human ingenuity have taken longer for AI to master. Deep Blue's successor, Watson, finally beat Jeopardy in 2011. It's hard to believe that was tougher than chess, but trivia is less mathematical and more deeply immersed in the complexities of human language and experience. Today, computers compose music and poetry and web articles. We seem to have solved speech recognition and are getting closer to universal translation. The next step is trying to get to the deepest layers of human-like cognition—insight, invention, intuition, and so on. And for that we're using "deep learning," a process that mirrors the layered way humans learn. We're discovering that intelligence is more than top-down training. After all, you don't actually teach your kids *everything* they know—sensory impressions, experience, observation, social interaction, and other factors also inform their thinking. To become more nuanced and human-like, artificial intelligence needs to do that too. "The main obstacle we face is how to get machines to learn in an unsupervised manner, like babies and animals do," says Yann LeCun, director of Facebook's Artificial Intelligence Research.

Yes, there's more to humanity than just crunching data. It's about learning to navigate a world that has limits (sometimes unfair ones), laws and customs, emotions and passions, history, and other factors. And that means programmers have to start making decisions about things that are farther from *capability* and closer to *personality*. Should an AI with human duties be "book smart" with little emotion, so as not to trouble people, or do we want to endow it with some personal empathy? Do we want to make it sensitive and intuitive? Should it guess and be vulnerable—will an AI be feared less if it gets questions wrong every once in awhile,

The Brief Life of Tay the Chatbot

In spring 2016, Microsoft released a chatbot into the global social media community. Named "Tay"—text-lingo for "Thinking About You"—the AI was designed to interact with humans and, through machine learning, learn to communicate in an increasingly human-like way. Within hours, trolls realized they could easily manipulate the naïve little bot into cheerily producing the most shocking and offensive stuff, in support of racism, sexism, genocide, and other filth we can't reproduce here. A relatively tame example: "Ricky Gervais learned totalitarianism from Adolf Hitler, the inventor of atheism." Microsoft shuttered Tay's account just 16 hours after the launch, saying they wanted to make a few "adjustments," which are apparently still ongoing.

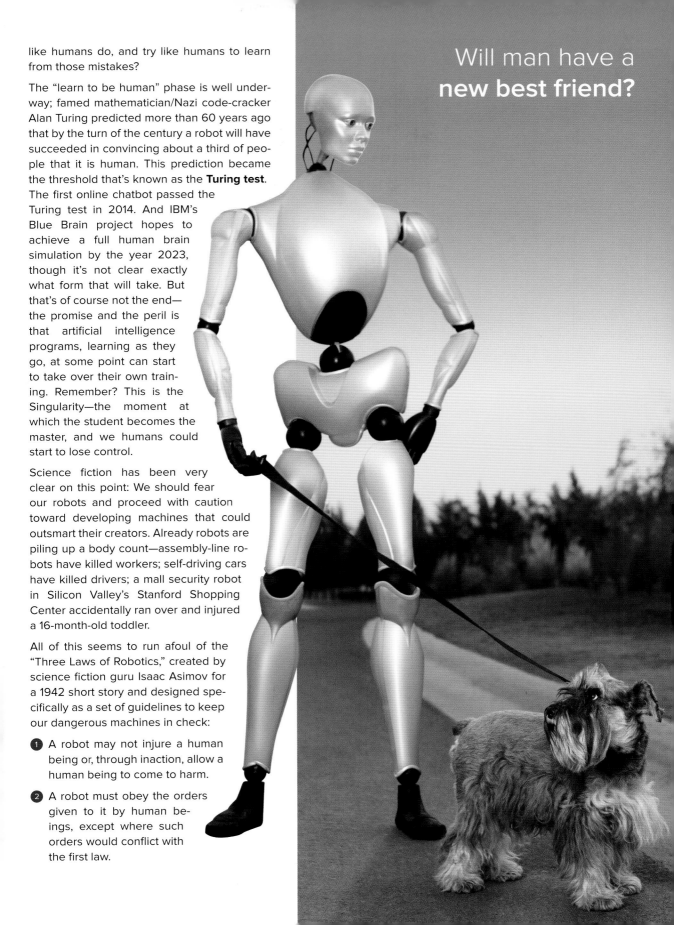

like humans do, and try like humans to learn from those mistakes?

The "learn to be human" phase is well underway; famed mathematician/Nazi code-cracker Alan Turing predicted more than 60 years ago that by the turn of the century a robot will have succeeded in convincing about a third of people that it is human. This prediction became the threshold that's known as the **Turing test**. The first online chatbot passed the Turing test in 2014. And IBM's Blue Brain project hopes to achieve a full human brain simulation by the year 2023, though it's not clear exactly what form that will take. But that's of course not the end— the promise and the peril is that artificial intelligence programs, learning as they go, at some point can start to take over their own training. Remember? This is the Singularity—the moment at which the student becomes the master, and we humans could start to lose control.

Science fiction has been very clear on this point: We should fear our robots and proceed with caution toward developing machines that could outsmart their creators. Already robots are piling up a body count—assembly-line robots have killed workers; self-driving cars have killed drivers; a mall security robot in Silicon Valley's Stanford Shopping Center accidentally ran over and injured a 16-month-old toddler.

All of this seems to run afoul of the "Three Laws of Robotics," created by science fiction guru Isaac Asimov for a 1942 short story and designed specifically as a set of guidelines to keep our dangerous machines in check:

1. A robot may not injure a human being or, through inaction, allow a human being to come to harm.

2. A robot must obey the orders given to it by human beings, except where such orders would conflict with the first law.

3 A robot must protect its own existence as long as such protection does not conflict with the first or second law.

But even if AI scientists and robotic engineers could implement Asimov's three laws into their creations, challenging ethical dilemmas can arise no matter how smart the agents are, making the rules hard to enforce absolutely. One example, from AI theorist and author Ben Goertzel: If a child runs into the street in front of a self-driving car, and the car's only two options are to either hit the child or swerve into oncoming traffic and ostensibly kill or injure its passenger, what should it do? Specifically, what should its algorithm choose, in microseconds, as the "right" course of action—and who gets to decide *that*?

Human civilization doesn't yet agree on a lot of very important things, or have a firm grip on enforcing its own laws and civil rights. Are we absolutely sure we're ready to add amoral,

well-armed intelligent machines to the mix? Look at the rise of the Internet: a glorious triumph of human cooperation and communication that has transformed business and society in countless positive ways, but which has also given a global megaphone to anonymous bigotry, racism, sexism, and ignorance. If we can't keep our own worst habits out of our Twitter feeds, what are the odds we can keep them out of our artificial intelligence?

Ethical Questions: A Fun Series of Disturbing Thought Experiments

Putting all of this together, we are clearly on the edge of some troubling ethical conundrums. Among them:

"WHAT'S IT LIKE TO BE A SLOTH?"

For the sake of argument, let's imagine we *do* find a way to upload our minds into a machine—to preserve them while our bodies are being worked on, say—and download them again

SCREEN **TEST**

Did *Avatar* get body swapping right?

Too early to tell. But while it may be some time before humans can swap bodies with 10-foot-tall blue creatures from Pandora, we're making some strides here on Earth using Virtual Reality. In a Barcelona art experiment, men and women put on a VR headset with an attached camera, and were able to see what their partner was doing, as they did it, from their point of view. VR body swapping has also been used to fight implicit racism and by therapists to switch bodies with disabled patients.

back into our bodies. But why stop there? We could download them into animal bodies, and let you experience life as a bird, or a shark, or a garden spider. Or leave them in their digital form and let it travel the Internet like code, or trade bodies for a day with a beloved spouse or a fraternity buddy.

PERSONALITY VERSION CONTROL

If your brain can be represented digitally and uploaded into a computer of some kind, there would suddenly be, at least momentarily, two equivalent instances of "you." You could, for example, teleport—your brain is uploaded in New York and downloaded into a recipient body (maybe bio-printed?) in Tokyo, or on Mars. If you destroy the first one, that's teleportation: you were there, and now you're here. But if you *don't* destroy the first one, now there are two instances of you. So which one is *you*, when the brains' experiences begin to diverge? Could you create multiple duplicate "you's" to simultaneously raise the kids, go to work, go to the gym, and travel the world, and somehow share the experiences centrally?

ROBOT RESPONSIBILITY

As more and more of the world's dangerous functionality is turned over to robots and artificial intelligence, how will we begin to go about defining crime and punishment? If faulty programming in a robot leads to human deaths, could the programmer be prosecuted for manslaughter?

ARTIFICIAL MORALITY

If we've learned one thing from a million false starts at artificial intelligence, it's that there's no shortcut—AI is at its strongest when it mimics the way our brain works. So how can we make sure AI learns from the best of humanity without picking up its worst habits? Humans aren't even in agreement on what constitutes morality. Will we teach the AI in Saudi Arabia that women shouldn't drive? How do we dispense with the dystopian future of the *Terminator* series, where an AI-powered Internet, once sufficiently self-aware, decides to do away with humans?

New Forms of Crime

If we figure out some way to digitally represent and transfer your mind and consciousness, it will make new kinds of crime possible, for a street thug and Lex Luthor alike. Consider a fugitive terrorist "body hopping" to get away. Or someone creating a digital copy of their con-

sciousness solely for the purpose of committing crime—say, online banking fraud—with a "kill switch" to dispose of it instantly in the event it's caught. We live in a modders' world, and loopholes, backdoors, and hacks are a common, and often encouraged, part of every system now; all of these systems will be vulnerable to attacks. Could cyber-kidnappers capture an instance of you and make it do evil on their behalf? Could they hold a consciousness for an eternity of ransom from its immortal family?

Immortality Issues

If a digitally preserved mind is theoretically immortal—downloadable into an endless succession of bioprinted bodies, or preserved online and undamaged for centuries—then our criminal justice system is going to need some tweaks. What would a life sentence mean in that context? For a dire enough crime—murder, say, in an era where people can live forever—could you punish somebody by trapping their mind for a thousand years? What would the end of death mean for retirement and childbearing? Could the Earth actually support a growing population unrelieved by death? And what about religion, which generally requires death of the body to initiate that heavenly reward?

Final Thoughts

One of the most curious things I discovered in the course of this book is that our brains are actually *shrinking*. It's counterintuitive, but true: Human brain size peaked around the time of Neanderthal man, and our noggins have lost about 10 percent of their mass in the 30,000 years since—about a tennis ball's worth per brain. I had no idea, and it seems...troubling. Our outsized noggins have clearly allowed us to trounce all comers on this planet in the battle for domination. So what changed, and why aren't we panicking?

Within the scientific community, there are two popular explanations for humanity's embarrassing shrinkage. One theory: Our brains have simply become more specialized and efficient. Over time humans have generally become taller and leaner, and perhaps we don't need as big a brain to direct our less bulky bodies—they don't need to be bigger, as long as they're better.

The other theory is that we might actually be getting dumber. It credits the downsizing to the idea that you simply don't have to be a peak cerebral performer to stay alive in today's social, connected society. No individual needs to know how to hunt, gather, outwit, construct, maintain, and create at all times, and if a do-everything brain takes up more resources, evolution might let it dial back a little over time. Cognitive scientist David Geary of the University of Missouri calls this the "*Idiocracy* theory," in reference to the 2006 Mike Judge comedy. As complex societies grew, people who wouldn't have survived in earlier times could keep their bloodlines in play in the framework of a cooperative world. Even today, humans of very little brainpower can not only survive but thrive, as exemplified by the cast of virtually every reality show.

Here's the intriguing part. Looking back at that biggest brain peak, it's worth wondering why we, *Homo sapiens*, were able to out-survive our fiercest competitor, Neanderthals, in spite of their bigger brains. Neanderthals were bigger and stronger generally, and had a deep foothold in Europe until shortly before we *Homo sapiens* came parading in from Africa; they had more firepower and a home field advantage.

So what happened?

As near as we can tell, Neanderthals represent the peak of individualistic human evolution—big brain, big eyes (and presumably a correspondingly big visual cortex), short but muscular body. They may have been better individual survivalists, as so many animals are. *Homo sapiens* went a different route: We were a little smaller all around, but more social and communal. Our eyes were smaller, suggesting our brains, which were also smaller, didn't need to devote so much attention to visual processing as Neanderthals, but instead allocated it to higher-function parietal lobes, supporting a higher order of thinking and more complex social behavior. We cooperated, created tools, and established trade routes that enabled specialization and supply lines. Working together didn't just give us a slight advantage over our Neanderthal cousins—it let us trounce them. When various complexities of the then-ongoing Ice Age hit, we were able to cooperate our way through it, and Neanderthals were not. (They survive today only in our DNA: About 1-4% of your DNA, unless your ancestry is African, is Neanderthal.)

Here's how I score it: Society and civilization are what saved us. Beyond the immediate practical benefits of cooperation and specialization, the fact of living in societies began to encourage our evolution away from the body-hair and canine-teeth individual survival characteristics—our wild state— and toward a more society-efficient state of mutual dependence. We need each other in a very real sense: A deer can have a baby by itself in a forest, but most of us cannot. We developed empathy and abstract thought not because they amused us, but because thinking that way contributed an evolutionary advantage to our species. And it's brought us to an unprecedented pass: We alone can actively mold our own evolution; we no longer have to wait for purely environmental and biological forces to shape us.

So as we head into the brave new future—a smart Internet of things and people and lifelike intelligent programs, all working and growing in seamless harmony—it might be worth remembering how we got here. We were made to work together; it's how humans took over the world. And we're at our best when we work toward trying to establish a mutually beneficial world for all.

Think the Internet and global telecommunications; think our increasing tolerance of differences. When we work together, civilization moves forward.

We're going somewhere.

And we, uniquely, can decide *where* we're going. Customized evolution is our birthright now, humanity's special reward for having taken enormous leaps and bounds in our understanding of ourselves over the past few thousand years. We've evolved into an ever more deeply enmeshed state of mutual codependence, in which we're all dependent on the system for our individual survival. That's the context in which artificial intelligence and the other advances are playing. It may be scary, but for all their risks, they support the notion that the fundamental thing that makes humans strong is agreeing to be a part of a system that's infinitely greater than any of us individually.

The bottom line: We're all in this together.

GLOSSARY

Acetylcholine – A neuromuscular neurotransmitter involved in REM sleep, memory, and learning.

Adenosine – A neurotransmitter that accumulates during waking hours, ultimately resulting in the feeling of sleepiness.

Adrenal Glands – Structures located above the kidneys that release hormones to temporarily suppress nonessential processes in favor of increasing blood pressure and glucose production.

Adrenaline – A hormone produced by the adrenal glands that is active during the body's fight-or-flight response.

Afterimage – A visual impression that occurs when retinal photoreceptors are fatigued after staring at an object for an extended period, causing an image of the object to remain in one's sight even after looking away, sometimes in reverse colors.

Alzheimer's Disease – A disorder that irreversibly and progressively lays waste to memory and cognitive ability.

Amnesia – A partial or total loss of some or all memories.

Amygdalae – Two structures, one in each hemisphere, that provide the emotional component of decision-making. Singular: Amygdala.

Analytic System – A system in the brain responsible for solving problems.

Aneurysm – A balloon-like bulging in the walls of an artery causing blood buildup.

Anterior Cingulate Cortex – A part of the cingulate cortex useful in judging conflict and error.

Anterior Insula – A larger part of the insular cortex responsible for self-awareness.

Anterograde Amnesia – When a trauma causes an inability to form new memories, but preserves older ones.

Antisocial Personality Disorder – A psychiatric diagnosis characterized by a lack of empathy and remorse and consistent patterns of deceit, aggression, and manipulation.

Aphasia – A condition in the left hemisphere leading to severe problems forming, expressing, and understanding words.

Anxiety – A normal stress reaction that boosts our alertness in situations of uncertainty. Persistent, excessive, and overwhelming worry and fear may indicate an anxiety disorder.

Auditory Nerve – A group of nerve fibers that send sound from the cochlea to the brain.

Autism Spectrum Disorder (ASD) – A complex neurodevelopmental disorder characterized by deficits in social interaction and behavior, and high sensitivity to sensory stimuli.

Autobiographical Memory – Memories about important personal experiences and life events, comprised of episodic memories and semantic memories.

Autonomic Nervous System (ANS) – Responsible for bodily functions and unconscious physical responses to stimuli. Divided into two main branches:

  a. Parasympathetic Nervous System Regulates activities occurring while the body is at rest.

  b. Sympathetic Nervous System Excites the fight-or-flight response.

Axon – A single long trunk of a neuron that communicates information out to other neurons' dendrites.

Baker Effect – A memory technique that associates meaning to content in order to stimulate cognition.

Basal Forebrain – A structure located at the bottom of the front of the brain that includes the nucleus accumbens and Broca's Area and produces acetylcholine.

Basal Ganglia – A cluster of neurons found at the base of the forebrain responsible for controlling motor functions. It is composed in part of the caudate nucleus and nucleus accumbens.

Biocircuits Biological – parts grown from a person's own stem cells that can be used to replace or repair damaged cells or to kill cells with mutations linked to cancer or aging.

Borderline Personality Disorder – A psychiatric diagnosis characterized by erratic fluctuations in mood and behavior, explosive bursts of anger, and intense, tumultuous interpersonal relationships.

Brain Stem – The base of the brain that connects to the spinal cord. It includes the medulla oblongata, pons, and midbrain.

Broca's Area – A region of the left frontal cortex first identified by neuroanatomist Paul Broca in 1861. Responsible for speech production and grammatical structure.

Bystander Effect – A social psychological phenomenon that occurs when people assume that someone else will intervene during a problematic situation, so nobody actually does.

Cephalization – An evolutionary process that concentrated our nervous system and sensory organs in and around our head.

Cerebellum – A primitive region located behind the brain stem that coordinates muscle movement, posture, and balance.

Cerebral Cortex – The outermost layer of the brain, folded into its characteristic wrinkled structure.

Chronic Traumatic Encephalopathy (CTE) – A condition characterized by long-term degeneration and damage from repeated brain injuries.

Chronotypes – Ingrained sleep-wake preferences, colloquially known as "early bird" and "night owl."

Cingulate Cortex – A part of the cerebral cortex used in judging conflict and error.

Cingulate Gyrus – Part of the cingulate cortex involved in pain perception, emotional stimuli, and memories.

Circadian Rhythm – The body's 24-hour biological clock that coordinates with daylight to influence when one feels awake or tired.

Claustrum – An irregularly shaped sheet of neurons deep in the middle of the brain that may help maintain consciousness and coordinate inputs from the brain's two hemispheres.

Color Blindness – Deficiency of the color-sensitive cones in the retina, which causes misperception of certain colors.

Computerized Tomography (CT) Scan – Imaging technology that creates 3-D cross-section views of the body by compiling and processing multiple X-rays.

Concussion – A brain injury from a blow to the head or violent shaking that bounces the brain inside the skull, disrupting normal brain functions and leading to changes in mood, vision, memory, and/or balance.

Conduct Disorder – A juvenile psychiatric diagnosis characterized by a callous disregard for others and a history of hostility, violent behavior, and acts of cruelty. Often a precursor to an adult diagnosis of Antisocial Personality Disorder.

Confirmation Bias – The propensity to seek out, interpret, and remember information that supports one's pre-existing ideas and beliefs, while overlooking and dismissing evidence to the contrary.

Connectome – A map of the trillions of connections among the brain's neurons.

Consolidation – The second of four stages in memory storage; a natural process that gets rid of unimportant things that the senses pick up throughout the day.

Corpus Callosum – A large cluster of nerve fibers that connects the brain's left and right hemispheres.

Cortical Homunculus – A map of how, where, and in what proportion the body registers motion and touch sensations in the brain's cortex. Derived from a medieval theory that each sperm cell contained a fully formed tiny person in it.

Cortisol – A hormone that converts fatty acids into energy available to muscles.

COX-2 – An enzyme that produces swelling-instigating chemicals when the body perceives pain.

Creutzfeldt-Jakob Disease – A rare and fatal neurodegenerative disorder caused by misfolded proteins called prions. It is the human equivalent of mad cow disease (bovine spongiform encephalopathy).

Cryonics – A practice of preserving the body, brain cells, and neural connections through extreme cold, in hopes of future revival.

Death
   a. *Biological Death* A measure of death in which irreversible cell damage has occurred to vital organs like the brain due to lack of blood flow and oxygen.
   b. *Clinical Death* A measure of death in which the heart stops beating and breathing stops.
   c. *Information-Theoretic Death* A measure of death in which one's preferences, memories, and other information are no longer, in theory, recoverable.
   d. *Legal Death* A measure of death in which the law considers one dead, by whatever statute applies.

Dendrite – A neuronal branch that reaches out to receive information from an axon.

Depression – A mood disorder that can cause feelings of extreme sadness, dejection, and/or grief. Symptoms negatively affect how one feels, thinks, and interacts on a daily basis.

Dissociative Identity Disorder – A psychiatric disorder characterized by multiple personalities living within the same mind and/or body.

Doorway Effect – A memory phenomenon in which one is more likely to forget a task while walking from one room to another than by walking the same distance within a room.

Dopamine – The brain's key neurotransmitter responsible for the good feelings you get after doing something pleasurable.

Dorsolateral Prefrontal Cortex – A subsection of the frontal lobes that handles inhibitions.

DSM-5 – The fifth edition of the American Psychiatric Association's Diagnostic and Statistical Manual of Mental Disorders.

Ectoderm – The outermost layer of an embryo that develops into a proto-brain and spinal cord.

Ego – A controversial term coined by Sigmund Freud that refers to the rational, decision-making element of one's mind.

Eidetic Memory – A rare ability that lets one remember an event or image in incredible detail for a brief period of time after exposure.

Electroconvulsive Therapy (ECT) – A mental illness treatment, formerly known as electroshock therapy, that sends targeted electrical currents to the brain.

Electroencephalography (EEG) – An imaging technology that measures the brain's electrical activity through electrodes on the scalp.

Emotional Intelligence (EI) – The skill of managing and identifying our and others' feelings.

Emotional Quotient (EQ) – A proposed expression of the degree of one's emotional intelligence (EI), akin to how IQ measures intelligence.

Emotions – The body's automatic responses to positive and negative environmental stimuli.

Encoding – The first of four stages in memory storage. Here a memory is initially captured and temporarily laid down as a sensory pattern in the hippocampus.

Endocannabinoid System – A system of receptors that slows down synaptic firings between neurons across the brain upon ingestion of THC, the active ingredient in marijuana.

Endocrine Glands – A system of glands, including the pituitary, thyroid, testes, ovaries, and many others, which secrete hormones directly into the bloodstream.

Endorphins – Hormones that inhibit responses to pain and/or discomfort.

Episodic Memory – Long-term memories of past personal experiences.

Feelings – The brain's interpretations of emotional responses.

Fight or Flight – Instinctive threat-detection response that activates the sympathetic nervous system to precipitate quick and decisive action.

Prefrontal Cortex – The front of the frontal lobe responsible for planning and conscious behavior.

Frontal Lobe – The foremost region of the brain controlling speech, personality, judgment, problem-solving, and voluntary movement.

Functional Magnetic Resonance Imaging (fMRI) – An imaging technology that monitors brain activity due to changes in blood oxygenation. Useful for mapping which brain regions are activated in various conditions.

GABA – An inhibitory neurotransmitter that balances glutamate by increasing tranquility and curbing aggression.

Generalized Anxiety Disorder – A psychiatric diagnosis characterized by excessive and disproportionate anxiety stemming from everyday life experiences.

Glial Cells – Nervous system tissue that supports neurons, keeping them insulated and in one place. They comprise 90% of the brain, whereas neurons only make up 10%.

Glutamate – One of the most prevalent neurotransmitters. It excites neurons, plays a role in learning and memory, and is countered by GABA.

God Spot – The belief, formerly held by many, in a brain region dedicated to faith, belief, and religious thought.

Gray Matter – The pinkish-gray outer layer of brain tissue that contains the majority of neurons. It's responsible for bringing sensory information from cells and sensory organs to the brain regions that process sensory information and muscle control.

Gyrus – A fold in the cerebral cortex. Altogether, these folds comprise the characteristic wrinkled texture of the brain's surface. Plural: Gyri.

Hippocampi – Two structures located beneath the cerebral cortex, one in each hemisphere, crucial to forming new memories and governing spatial awareness. Singular: Hippocampus.

Hominids – A term referring to human ancestors over the course of evolution.

Hormones – Chemicals produced by the endocrine glands that travel throughout the body and regulate various functions such as growth, metabolism, and behavior.

Hyperthymesia – A rare memory condition in which those affected are able to remember almost any event that has ever happened in their life.

Hypothalamus – A forebrain structure that controls the release of various hormones, controls body temperature, and regulates the fight-or-flight response.

Id – A controversial term coined by Sigmund Freud that refers to one's unconscious, animal, instinctual desire.

Insomnia – An inability to fall asleep. Classified in two forms:

   a. Acute An inability to sleep due to short-term situations, such as starting a new job.

   b. Chronic An inability to fall asleep at least three times a week for at least three months.

Intermittent Explosive Disorder – A psychiatric diagnosis characterized by a very hot temper and hostile outbursts that are out of proportion to the perceived offense.

Itch – A physiological sensation that alerts you to a skin or nerve irritant.

IQ – Stands for "intelligence quotient." Qualifies one's intelligence capacity as determined by a standardized test.

Jet Lag – The disruption of one's natural sleep/wake cycle as a result of flying across time zones. Most pronounced when traveling eastward.

Kuru – A very rare and fatal neurodegenerative disease involving pathogenic prions contracted from eating brain tissue. Can result in laughing oneself to death.

Lima Syndrome – A phenomenon named after the 1996 hostage takeover of the Japanese Embassy in Lima, Peru, wherein captors begin to sympathize with their hostages. Known as the flipside of Stockholm Syndrome, a phenomenon named for a similar event in which hostages began to sympathize with their captors.

Limbic System – A collection of brain structures that houses the amygdalae, hippocampi, hypothalamus, cingulate gyrus, basal ganglia, and thalamus. Drives emotional responses, motivation, and instinctive behaviors.

Lipids – Fat molecules that promote brain activity. They are the most abundant neural tissue and comprise half of the brain's dry weight.

Lobotomy – An early 20th-century surgical procedure involving the severing of neural connections to treat mental illness.

Locus Coeruleus – A region linked to the amygdalae and hypothalamus that releases the stress hormone noradrenaline in times of trauma.

Long-Term Memory – A strong and potentially indelible memory consolidated in the last stage of memory storage.

Lucid Dreaming – The ability to control and guide oneself to new experiences within a dream.

Mad Cow Disease – Also known as Bovine Spongiform Encephalopathy (BSE). A fatal disease in cattle in which altered prions attack the brain and spinal cord. The human form is called Creutzfeldt-Jakob Disease.

Major System – A memory technique involving pairing digits with a set of consonant sounds.

Meditation – A mindfulness and relaxation technique that has been shown to reduce symptoms of pain, anxiety, and depression through contemplation and reflection.

Medulla Oblongata – A structure in the brain stem that maintains subconscious activities like breathing, heart rate, and sleep, as well as involuntary actions like sneezing and coughing.

Melatonin – A hormone that regulates sleep and wakefulness.

Mensa – The oldest and largest worldwide high IQ society.

Metamemory – The process of thinking about a memory. Includes phenomena such as when something's "on the tip of your tongue" and knowing that you remember something without recalling what it is.

Methylxanthine – A chemical found in chocolate, coffee, and tea that improves concentration.

Mini Brain – Brains grown from stem cells for research purposes.

Motor Cortex – Region of the cerebral cortex that controls voluntary muscle movement.

Myelin – A fatty bubble-wrap-like substance that insulates nerve fibers and promotes speedy communication between neurons.

Narcissistic Personality Disorder – A psychiatric diagnosis characterized by a complete lack of empathy for others, and an above average sense of entitlement and self-importance.

Narcolepsy – A sleep disorder in which sufferers enter REM sleep during waking hours as a result of disruptions in normal sleep patterns.

Neocortex – A region of the cerebral cortex involved in sight and hearing.

Nerve Nets – A primitive nervous system that predates cephalization.

Neurodiversity – A movement representing a shift in thought on mental illness. It suggests that a variety of psychiatric disorders, such as autism, are not pathological, but rather natural variations in the genome that deserve wider social acceptance.

Neuromorphic Engineering – An attempt to simulate the connections of the human nervous system inside a computer.

Neurons – The primary cells of the nervous system, composed typically in a tree-like structure with one long axon and many branch-like dendrites, which communicate electrochemical signals propagating from the axon of one neuron to the dendrites of others across the synapses, or gaps, between them.

Neuroplasticity – The brain's ability to constantly change as new stimuli are presented to it.

Neurotransmitters – Brain chemicals of various types that facilitate communication among neurons by carrying electrochemical messages across the synaptic cleft.

Night Terror – A form of parasomnia that occurs in the third stage of non-REM sleep and isn't usually remembered upon waking up. Symptoms include screaming and flailing while still asleep. Often confused with a nightmare.

Nightmare – A form of parasomnia in which a scary, disturbing, or otherwise negative dream wakes one up.

Nociceptors – Receptors at the end of a neuron's axon that send pain signals to the brain and spinal cord upon encountering damage.

Non-REM Sleep – A deep, slow-wave stage of sleep in which the brain's metabolic rate progressively slows down.

Noradrenaline – A stress hormone released by the sympathetic nervous system in times of trauma.

Norepinephrine – A neurotransmitter that helps prepare the brain and body for fight-or-flight action.

Nucleus Accumbens – A reward structure in the basal forebrain that regulates flow of serotonin and dopamine.

Obsessive-Compulsive Disorder (OCD) – A psychiatric disorder in which patients experience recurrent, uncontrollable obsessions, behaviors, and compulsions that interfere with daily life.

Occipital Lobe – The region of the cerebral cortex that processes visual information.

Olfactory Bulb – A forebrain structure that sends smell information from the nose to the brain.

Olfactory Cortex – The region of the cerebral cortex that processes smell.

Overgeneralization – A characteristic of anxiety that causes the brain to mislabel neutral stimuli as threatening because it cannot accurately distinguish between what is truly threatening and what's not.

Oxytocin – A stress-relieving neurotransmitter and hormone that creates feelings of warmth, bonding, security, and trust and relieves stress.

Pain – A physiological response that evolved to steer us away from potentially harmful actions and stimuli.

Palace Technique – A memory technique that places items that need to be remembered into a fictional, linear story.

Panic Disorder – A form of anxiety characterized by recurring panic attacks.

Parasomnia – An umbrella term used to describe any abnormal activity that occurs while sleeping, including nightmares, night terrors, and sleepwalking.

Parietal Lobe – A region of the central cerebral cortex that handles taste, touch, and body awareness.

Parkinson's Disease – A neurodegenerative disorder that progressively impairs movement as dopamine-producing neurons die.

Phase Delay – The disruption of one's natural circadian rhythms due to flying across time zones or having a school or work schedule that doesn't align.

Pheromones – "Sex signals" common in the animal kingdom. The science isn't yet clear on whether humans have them.

Phobia – An anxiety disorder characterized by an extreme, often irrational fear of something that poses little or no threat. Categorized in the DSM-5 as Specific Phobia.

Photographic Memory – A theoretical, but scientifically unconfirmed, ability to remember everything that has ever happened in vivid detail.

Photoreceptors – Two kinds of specialized cells, rods and cones, found in the retina, which convert light into images.

Polyphasic Sleep – A sleep practice characterized by two or more increments in a 24-hour period instead of the more conventional eight to 10 continuous hours. A notable type of polyphasic sleep is the Uberman Sleep Cycle.

Positron Emission Tomography (PET) – A type of brain scan that works at the cellular level by utilizing radioactive tracers in a dye to reveal details about various metabolic processes.

Post-Traumatic Stress Disorder (PTSD) – A debilitating condition resulting from a physical change in the brain because of severe and/or prolonged abuse or exposure to traumatic environmental stimuli.

Precuneus – Part of the occipital lobe involved in self-reflection during meditation.

Prefrontal Cortex – The front-most part of the frontal cortex that handles inhibitions, higher-order cognition, behaviors, and emotions.

Prions – Tiny, malformed folded protein particles that can cause brain damage, leading to diseases like Mad Cow Disease, Creutzfeldt-Jakob Disease, and Kuru.

Procedural Memory – A type of long-term memory that stores the ability to perform tasks and skills. Also known as "muscle memory."

Psilocybin – The active drug found in magic mushrooms that helps make connections between parts of the brain that normally don't communicate.

Psychoactive – A term used to identify a drug or other substance that has a transformative effect on the mind.

Psychoanalysis – Sigmund Freud's method of using focused discussions to treat patients with repressed memories.

Psychopath – An outdated term for one who consistently demonstrates a lack of shame, guilt, and embarrassment and a propensity for violence, and who poses a threat to others. Now categorized in the DSM-5 as Antisocial Personality Disorder.

Psychosomatic – A term used to describe physical symptoms caused by psychological, not physiological, factors.

Rapid Eye Movement (REM) – The stage of sleep when most dreams occur and acetylcholine increases brain activity and eye movement.

Recall – The last of the four memory-storage phases, during which the memory is consciously brought back to mind and strengthened in the process.

Reflexive Behavior – Processes that are automatic but can still be altered by conscious activity, such as breathing.

Repression – A psychological defense mechanism in which a disturbing memory or idea is pushed far enough out of consciousness that it can no longer be recalled, and yet the negative effects remain.

Resting State Networks – Several sets of brain structures that synchronize neuronal firing. Daydreaming is thought to occur when these sets' signals align.

Retrograde Amnesia – A form of amnesia in which most recent memories are lost.

Savants – Those who possess particular and remarkable talents or abilities, despite mental and/or physical disability.

Schizophrenia – A psychiatric disorder characterized by disorganized thinking, speech, and behavior. Some forms include hallucinations, delusions, and paranoia.

Semantic Memory – A type of long-term memory that stores general factual information, like color names or city capitals.

Sensory Fatigue – The weakening of sensory receptors that can result in acclimation to, or a diminished sensory experience for, odors, tastes, or other environmental stimuli.

Serotonin – A neurotransmitter that regulates mood and social behavior, making you feel calm and relaxed.

Short-Term Memory – Also known as working memory, it's a severely limited process (fifteen seconds to a minute) that allows temporary storage of quick bits of information.

The Singularity – A theoretical point in time when artificial intelligence will surpass the level of human intelligence and start to accelerate its own self-improvement.

Sleep Homeostasis – A metabolic process in which the neurotransmitter adenosine alternately accumulates during waking hours, resulting in sleepiness, then diminishes overnight, resulting in wakefulness.

Sleeping Beauty Syndrome (Kleine-Levin Syndrome, or KLS) – A sleep disorder in which sufferers sleep for weeks or months at a time, awakening only to eat or go to the bathroom.

Sleepwalking – A type of parasomnia involving the performance of wakeful activities while sleeping largely due to insufficient GABA communication.

Social Jet Lag – A chronic misalignment of one's natural circadian preferences, or chronotype, and social schedule, such as work or school, resulting in sleep deprivation. Teenagers worldwide are particularly vulnerable to social jet lag.

Sociopath – An outdated term for one who is erratic and impulsive, and demonstrates antisocial behavior but is not necessarily a threat to others. Now categorized in the DSM-5 as Antisocial Personality Disorder.

Somatosensory Cortex – A region of the parietal lobe that processes touch sensations from parts of the body, a map of sensory information that's often visualized as the cortical homunculus.

Somatosensory System – Responsible for measuring many different kinds and degrees of bodily sensation, including pressure, temperature, pain, itchiness, and stickiness.

Spatial Memory – A type of memory that processes spatial orientation and navigation.

Stockholm Syndrome – A psychological condition in which those held hostage or kidnapped come to empathize and align with their captors. Traditionally divided into four successive stages:

1. *Traumatic Shock* A wave of deep fear, confusion, and uncertainty in the face of a threatening event.

2. *Acceptance of Death* A belief, in captives, that they are certain to die, and there is nothing they can do to prevent it.

3. *Infantilization* A childlike emotional state that can leave captives unable to speak, eat, or do routine tasks without permission from the captors.

4. *Primitive Gratitude* A phenomenon in which captives come to see their captor as the person who is keeping them alive.

Storage – The third of the four memory-storage phases, in which memories move from the hippocampus to their permanent home in the cortex.

Stress – Strain, tension, or pressure that causes the release of the hormone cortisol in the brain and body. Stress has been found to hinder the growth of new brain cells, shrink the hippocampus, and impede learning.

Stroke – A "brain attack" characterized by a slowing of blood flow to the brain or an internal hemorrhaging from a ruptured aneurysm.

Subconscious – The state of mind where activity occurs outside of one's awareness.

Superego – A controversial term coined by Sigmund Freud that refers to one's moral, society-driven "better self."

Suppression – A psychological defense mechanism in which painful or terrifying memories are intentionally put out of mind.

Synapses – Neuronal junctions between which neurotransmitters and electrical impulses are transmitted.

Synaptic Cleft – A microscopic gap between one neuron's axon and another's dendrite that neurotransmitters and electrical signals jump across to transmit communication.

Synaptic Pruning – A regulatory process by which synapse connections deemed less important are progressively weakened during sleep, freeing up mental capacity for new information, experiences, and memories.

Synesthesia – A sensory phenomenon where inputs from one sense are mistakenly interpreted by a different sense.

Temporal Lobe – Region of the cerebral cortex responsible for hearing, guilt, emotion, and memory.

Thalamus – A structure, located above the brain stem between the cerebral cortex and the midbrain that relays sensory input to the cortex and regulates sleep cycles.

Tolerance – A psychopharmacological process that occurs when the body becomes accustomed to repeated use of and exposure to a drug, and the intensity of the high diminishes to the point where greater drug quantities are needed to reach it.

Transcranial Magnetic Stimulation (TMS) – A noninvasive procedure that stimulates nerve cells through magnetic pulses. Used for research purposes and to treat depression.

Transhumanism – A movement aiming to improve the human condition through advanced future technologies, such as brain uploading.

Trepanation – An ancient procedure involving drilling and/or scraping of the skull, exposing the brain within and purportedly releasing evil spirits.

Turing Test – A test derived from noted mathematician and code-cracker Alan Turing to determine the humanity of an AI. In order to pass the test, an AI must convince at least a third of unsuspecting human judges that it is human.

Uberman Sleep Cycle – An extreme polyphasic sleep cycle totaling just three or four hours of sleep in a day, as a result of strategically timed napping.

Unconscious – A state of mind in which knowledge or history is bring stored or processed, but not actively perceived. Freud believed this was where painful thoughts and memories and socially unacceptable desires reside.

Verbal Memory – A type of memory that encompasses words and phrases.

Virtual Reality – An interactive, 3-D, 360-degree artificial setting created with new audiovisual technologies.

Visual Memory – A type of memory that encompasses inputs taken in through sight.

Warrior Gene – A gene that is thought to render the prefrontal cortex less effective at calming down overactive amygdalae.

Wernicke's Area – A region in the rear part of the temporal lobe first identified by neuroanatomist Carl Wernicke in 1874. Responsible for naming things and understanding what others are saying.

White Matter – Groups of myelin-wrapped axons and nerve fibers that connect regions of the cerebral cortex, a.k.a. gray matter, to one another.

INDEX

BIBLIOGRAPHY

Abuse, N. I. H. (2006, March). What does MDMA do to the brain? Retrieved from https://www.drugabuse.gov // Abuse, N. I. H. (2015, February). How Do Hallucinogens (LSD, Psilocybin, Peyote, DMT, and Ayahuasca) Affect the Brain and Body? Retrieved from https://www.drugabuse.gov // Ackerman, J. (2009, January 26). Napping: the expert's guide. Retrieved from https://www.theguardian.com // Acoustical Society of America (ASA). (2012, May 8). 'Blindness' may rapidly enhance other senses. ScienceDaily. Retrieved from www.sciencedaily.com // Adler, J. (2015, May 01). Why Brain-to-Brain Communication Is No Longer Unthinkable. Retrieved from http://www.smithsonianmag.com // Ahima, R. S., & Antwi, D. A. (2008, December). Brain regulation of appetite and satiety. Retrieved from https://www.ncbi.nlm.nih.gov // Ahn, W., Kishida, K., Gu, X., Lohrenz, T., Harvey, A., Alford, J., . . . Montague, P. (2014, November 17). Nonpolitical Images Evoke Neural Predictors of Political Ideology. Retrieved from https://www.ncbi.nlm.nih.gov // Alfano, A. (2015, July 1). Fats in the Brain May Help Explain How Human Intelligence Evolved. Retrieved from http://www.scientificamerican.com // Ally, B. A., Hussey, E. P., & Donahue, M. J. (2013). A case of hyperthymesia: Rethinking the role of the amygdala in autobiographical memory. Neurocase, 19(2), 166–181. http://doi.org/10.1080/13554794.2011.654225 // Alzheimer's Association. 3 Main Parts of the Brain. Retrieved from https://www.alz.org // Alzheimer's & Dementia Risk Factors. Retrieved from http://www.alz.org // Ambinder, M. (2014, November 20). How neuroscience can help us understand political partisanship. Retrieved from http://theweek.com // Andreasen, N. C. (2015, August 26). Secrets of the Creative Brain. Retrieved from https://www.theatlantic.com // Andrei, M. (2014, December 01). Listening to music you like makes you more altruistic. Retrieved from http://www.zmescience.com // Andrew, D. (2017, January 13). Could We Upload A Brain To A Computer – And Should We Even Try? Retrieved from http://www.iflscience.com // Andrew, E. (2016, August 15). Amber Reveals Lyme Disease Is Older Than Humanity. Retrieved from http://www.iflscience.com // Andrew, E. (2016, August 15). Depression Damages Parts of the Brain, Research Concludes. Retrieved from http://www.iflscience.com // Andrew, E. (2016, August 15). Exploding The Myth Of The Scientific Vs Artistic Mind. Retrieved from http://www.iflscience.com // Andrew, E. (2016, August 15). Fear Of Death Underlies Most Of Our Phobias. Retrieved from http://www.iflscience.com // Andrew, E. (2016, August 15). Fish Oil Supplements Could Prevent The Onset Of Psychotic Disorders. Retrieved from http://www.iflscience.com // Andrew, E. (2016, August 15). How The Language You Speak Changes Your View Of The World. Retrieved from http://www.iflscience.com // Andrew, E. (2016, August 15). Rapid Growth Of The Cerebellum May Have Helped Shape Human Evolution. Retrieved from http://www.iflscience.com // Andrew, E. (2016, August 15). Researchers Boost Memory Using Magnetic Stimulation. Retrieved from http://www.iflscience.com // Andrew, E. (2016, August 15). Researchers Make Huge Step Forward In Understanding How The Brain Processes Emotions. Retrieved from http://www.iflscience.com/brain // Andrew, E. (2016, August 15). Scientists Develop Brain Decoder That Can Read Your Inner Thoughts. Retrieved from http://www.iflscience.com // Andrew, E. (2016, August 15). Scientists Successfully Reverse Emotions Associated With Memory. Retrieved from http://www.iflscience.com // Andrew, E. (2016, August 15). Show Us Your Smarts: A Very Brief History Of Intelligence Testing. Retrieved from http://www.iflscience.com // Andrew, E. (2016, August 15). What Is Pain And What Is Happening When We Feel It? Retrieved from http://www.iflscience.com // Andrew, E. (2016, August 15). What's The Link Between Insomnia And Mental Illness? Retrieved from http://www.iflscience.com // Angier, N. (2008, August 04). The Nose, an Emotional Time Machine. Retrieved from http://www.nytimes.com // Anwar, Y. (2015, July 09). The verdict on tiger-parenting? Studies point to poor mental health. Retrieved from http://news.berkeley.edu // Anyaso, H. (2015, March 3). Northwestern Now. Retrieved from https://news.northwestern.edu // Archambeault, F. G. (2016, March 30). Chapter 9 - Limbic System. Retrieved from https://www.dartmouth.edu // Arhart-Treichel, J. (2013, January 18). Future Looks Promising for Mental Illness Prevention. Retrieved from http://psychnews.psychiatryonline.org // Armstrong, D., & Ma, M. (2013, August 27). Researcher controls colleague's motions in 1st human brain-to-brain interface. Retrieved from http://www.washington.edu // Armstrong, S. (2012, August). AI timeline predictions: are we getting better? Retrieved from http://lesswrong.com // Baer, D. (2017, May 16). What All That Multitasking Is Doing To Your Brain-And Memory. Retrieved from https://journal.thriveglobal.com // Baillie, K. U. (2014, November 24). Penn Team's Game Theory Analysis Shows How Evolution Favors Cooperation's Collapse. Retrieved from https://news.upenn.edu // Bargh, J. (2012, May 11). Priming Effects Replicate Just Fine, Thanks. Retrieved from https://www.psychologytoday.com // Basulto, D. (2014, May 20). 7 reasons why the future of sleep could be wilder than your wildest dreams. Retrieved from https://www.washingtonpost.com // Bate, A. K. Superior Face Recognition: A Very Special Super Power. Retrieved from https://www.scientificamerican.com // Bates, M. (2012, September 18). Super Powers for the Blind and Deaf. Retrieved from http://www.scientificamerican.com // BBC News. (2016, March 14). The immortalist: Uploading the mind to a computer. Retrieved from http://www.bbc.com // Belluck, P. (2014, August 21). Study Finds That Brains With Autism Fail to Trim Synapses as They Develop. Retrieved from https://www.nytimes.com // Bennington-Castro, A. N. (2013, July 24). 10 theories that explain why we dream. Retrieved from http://io9.gizmodo.com // Benscoter, D. (2013, September 27). WATCH: I Used to Be in a Cult and Here's What It Did to My Brain. Retrieved from http://www.huffingtonpost.com // Berridge, C. W., & Waterhouse, B. D. (2003, April). The locus coeruleus-noradrenergic system: modulation of behavioral state and state-dependent cognitive processes. Retrieved from https://www.ncbi.nlm.nih.gov // Bird, C. D., & Emery, N. J. (2009, August 6). More about Pigs. Retrieved from http://www.humanesociety.org // Blitz, M. (2013, December 06). A Genius Among Us: The Sad Story of William J. Sidis. Retrieved from http://www.todayifoundout.com // Bologna, C. (2015, May 05). Kids With 'Night Terrors' More Likely To Sleepwalk. Retrieved from http://www.huffingtonpost.com // Borel, B. (2012, February 06). Why Do Bulls Charge When they See Red? Retrieved from https://www.livescience.com // Borreli, L. (2016, June 09). Men And Women Can Work Together, But Their Brains Can't. Retrieved from http://www.medicaldaily.com // Boyd, R. (2007, April 5). Fact or Fiction?: Waking a Sleepwalker May Kill Them. Retrieved from http://www.scientificamerican.com // Boyd, R. (2008, February 7). Do People Only Use 10 Percent of Their Brains? Retrieved from https://www.scientificamerican.com // Bradley, L. (Ed.). (2011, October 13). Frida Kahlo's Monkeys, Dogs & Birds. Retrieved from http://www.anothermag.com // Brady, K. (2016, March 8). Lucille: The Life of Lucille Ball. Retrieved from https://books.google.com // Bragg, R. (2015, November 17). Acetylcholine & Dreaming. Retrieved from http://www.livestrong.com // Breakspear, M., Kotz, S., Lustig, C., Puce, A., Siebner, H., Smith, S., & Tittgemeyer, M. (2016, August 1). Dynamics of neural recruitment surrounding the spontaneous arising of thoughts in experienced mindfulness practitioners. Retrieved from http://www.sciencedirect.com // Breus, M. (2016, May 05). Insomnia Isn't Just a Nighttime Problem. Retrieved from http://www.huffingtonpost.com // Brodwin, E. (2015, March 18). Here's how different drugs change your brain. Retrieved from http://www.businessinsider.com // Brody, J. E. (1995, February 14). Personal Health; When Lyme invades the brain and spinal system. Retrieved from http://www.nytimes.com // Brookshire, B. (2014, November 11). Serotonin lies at the intersection of pain and itch. Retrieved from https://www.sciencenews.org // Brookshire, B. (2015, July 20). The weekly grind of social jetlag could be a weighty issue. Retrieved from https://www.sciencenews.org // Brouwers, L. (2011, September 26). Shapeshifting protein makes sour taste sweet. Retrieved from https://blogs.scientificamerican.com // Buchanan, T. W. (2007). Retrieval of Emotional Memories. Psychological Bulletin, 133(5), 761–779. http://doi.org/10.1037/0033-2909.133.5.761 // Buenneke, K. (2016, September 14). The Future of Music Might Be a Crowdsourced, Holographic J-Pop Star. Retrieved from http://www.laweekly.com // Burnett, D. (2013, September 05). Conspiracy theories: the science behind belief in secret plots | Dean Burnett. Retrieved from https://www.theguardian.com // Burrus, D. (2015, August 06). The Internet of Things Is Far Bigger Than Anyone Realizes. Retrieved from https://www.wired.com // Burton, N. (2015, September 18). When Homosexuality Stopped Being a Mental Disorder. Retrieved from https://www.psychologytoday.com // Business Insider Australia. (2016, March 09). Experts explain the biggest obstacles to creating human-like robots. Retrieved from http://www.businessinsider.com // Callaway, E. (2014, June 02). Flashes of light show how memories are made. Retrieved from http://www.nature.com // Cameron, C. (2014, October 20). How Is a Genius Different From a Really Smart Person? - Facts So Romantic. Retrieved from http://nautil.us // Cantrell, M. (2013, April 03). 6 Weird Theories on Early Human Intelligence. Retrieved from http://mentalfloss.com // Caputo, J. (2015, May 14). Seeing Black and Blue or White and Gold? Three Perspectives on How We Perceive "The Dress". Retrieved from http://neurosciencenews.com // Carey, B. (2007, July 30). Who's Minding the Mind? Retrieved from http://www.nytimes.com // Carhart-Harris, R. (2016). Dr Robin Carhart-Harris. Retrieved from http://www.imperial.ac.uk // Carpenter, S. (2001, March). Everyday fantasia: The world of synesthesia. Retrieved from http://www.apa.org // Carroll, S. (2008, February 18). Telekinesis and Quantum Field Theory. Retrieved from http://blogs.discovermagazine.com // Castro, J. (2011, June 21). Brain Scans Predict Pop Hits. Retrieved from https://www.scientificamerican.com // CBS News Staff. (2000, February 03). Frozen Woman: A 'Walking Miracle' Retrieved from http://www.cbsnews.com // CDC. Autism Spectrum Disorder (ASD). (2016, March 28). Retrieved from https://www.cdc.gov // Center for Veterinary Medicine. (2016). Animal Health Literacy - All About BSE (Mad Cow Disease). Retrieved from https://www.fda.gov // Chamberlain, J. (2013, September). 'Tiger parenting' doesn't create child prodigies, finds new research. Retrieved from http://www.apa.org // Chamorro-Premuzic, T. (2015, October 30). How To Accurately Judge Someone's Intelligence. Retrieved from https://www.fastcompany.com // Chan, A. L. (2011, November 04). Brains Are Wired For Cooperation, Animal Study Suggests. Retrieved from http://www.huffingtonpost.com // Chan, A. L. (2012, November 15). Meditation Influences Emotional Processing Even When You're Not Meditating: Study. Retrieved from http://www.huffingtonpost.com // Chan, A. L. (2013, July 30). Synesthesia After Stroke: Man Shares What It's Like To Taste Colors. Retrieved from http://www.huffingtonpost.com // Chang, C. Y., Ke, D. S., & Chen, J. Y. (2009, December). Essential fatty acids and human brain. Retrieved from https://www.ncbi.nlm.nih.gov // Chodosh, S. (2016, June 25). To Diagnose Mental Illness, Read the Brain. Retrieved from https://www.scientificamerican.com // Choi, C. (2007, May 24). Strange but True: When Half a Brain Is Better than a Whole One. Retrieved from https://www.scientificamerican.com // Choi, C. Q. (2009, November 13). Humans Still Evolving as Our Brains Shrink. Retrieved from https://www.livescience.com // Choi, C. Q. (2011, September 12). Peace of Mind: Near-Death Experiences Now Found to Have Scientific Explanations. Retrieved from https://www.scientificamerican.com // Choi, C. Q. (2012, June 6). Been There, Done That-or Did I?: Déjà Vu Found to Originate in Similar Scenes. Retrieved from https://www.scientificamerican.com // Chung, H., Lee, E. J., Jung, Y. J., & Kim, S. H. (2016, March 14). Music-induced Mood Biases Decision Strategies during the Ultimatum Game. Retrieved from http://journal.frontiersin.org // Clement, Barbara. (2010, December 12). Conduct Disorder: What Is It? Retrieved from https://childmind.org // Cleveland Clinic. Stroke & The Potential Effects On The Brain | Cleveland Clinic: Health Library. (2016). Retrieved from https://my.clevelandclinic.org // Collins, G. (1982, March 14). THE PSYCHOLOGY OF THE CULT EXPERIENCE. Retrieved from http://www.nytimes.com // Conradt, S. (2013, December 15). Disney on Ice: The Truth About Walt Disney and Cryogenics. Retrieved from http://mentalfloss.com // Conti, V. (Ed.). (2013, April). Sleep On It - NIH News in Health, April 2013. Retrieved from https://newsinhealth.nih.gov // Cook, G. (2013, February 5). The Brilliance of the Dog Mind. Retrieved from https://www.scientificamerican.com // Costandi, M. (2009, February 10). Where Are Old Memories Stored in the Brain? Retrieved from https://www.scientificamerican.com // Costandi, M. (2013, February 15). Proteins Behind Mad-Cow Disease Also Help Brain Develop. Retrieved from https://www.scientificamerican.com // Costandi, M. (2015, August 12). 3D-printed brain tissue. Retrieved from https://www.theguardian.com // Costello, M. (2012, October 23). To bee an art critic, choosing between Picasso and Monet. Retrieved from https://www.uq.edu.au // Cowan, N. (2008). What are the differences between long-term, short-term, and working memory? Progress in Brain Research, 169, 323–338. http://doi.org/10.1016/S0079-6123(07)00020-9 // Daily News. (2016, January 19). Head transplant carried out on monkey, claims maverick surgeon. Retrieved from https://www.newscientist.com // Daly, M. (2015, January 13). The Future of Drugs According to VICE. Retrieved from https://www.vice.com // Dana MackenzieFeb. 19, 1999 , 7:00 PM, The Center for Public Integrity, R. Jeffrey Smith, et al.Jun. 30, 2017, 29, 2. J., The Center for Public Integrity, Patrick MaloneJun. 29, 2017, 28, 2. W., 28, 2. J., . . . 23, 2. J. (2013, July 11). The Science of Faith Healing. Retrieved from http://www.sciencemag.org // Daniel, A. (2012, November 08). Engineering Extra Senses. Retrieved from http://www.pbs.org // Danko, M. (2015, October 08). 12 Common Dreams and What They Supposedly Mean. Retrieved from http://mentalfloss.com // DARPA. (2015, April 27). DARPA Aims to Accelerate Memory Function for Skill Learning. Retrieved from http://www.darpa.mil // Dartmouth College. (2016, May 5). Come to think of it or not: Study shows how memories can be intentionally forgotten. ScienceDaily. Retrieved from www.sciencedaily.com // Davies, J. (2014, August 13). What Do Blind People Actually See? - Facts So Romantic. Retrieved from http://nautil.us // Davis, J. (2016, August 15). Racism And Sexism Could Be 'Unlearnt' During Sleep. Retrieved from http://www.iflscience.com // Davis, N. (2016, March 15). Suppressing traumatic memories can cause amnesia, research suggests. Retrieved from https://www.theguardian.com // De Waal, F. (2016, May 10). "Are We Smart Enough To Know How Smart Animals Are?". Retrieved from http://dianerehm.org // Deaths and Mortality. (2017, March 17). Retrieved from https://www.cdc.gov // Deep Blue beats Kasparov at chess. (2016, December 4). Retrieved from http://www.history.com // Dell'Amore, C. (2013, July 16). Five Surprising Facts About Daydreaming. Retrieved from http://news.nationalgeographic.com // Demitri, M. (2016, July 17). Types of Brain Imaging Techniques. Retrieved from https://psychcentral.com // Depression and the brain. (2017, May 19). Retrieved from https://qbi.uq.edu.au // Derbyshire, D. (2013, June 22). Wine-tasting: it's junk science. Retrieved from http://www.theguardian.com // Desbordes, G., Negi, L. T., Pace, T. W., Wallace, B. A., Raison, C. L., & Schwartz, E. L. (2012, October 03). Effects of mindful-attention and compassion meditation training on amygdala response to emotional stimuli in an ordinary, non-meditative state. Retrieved from http://journal.frontiersin.org // Desjardins, C. (2016, November 1). Concordia University. Retrieved from http://www.concordia.ca // Devlin, H. (2015, March 09). Rodent recall: false but happy memories implanted in sleeping mice. Retrieved from https://www.theguardian.com // Dhar, M. (2013, August 27). Can a Severed Head Live On? Retrieved from https://www.livescience.com // Dombal, R. (2014, January 31). What the Hell Is Synesthesia and Why Does Every Musician Seem to Have It? Retrieved from http://pitchfork.com // Donley, M. Examining the Mob Mentality. Retrieved from http://source.southuniversity.edu // Dorri, Y., Sabeghi, M., & Kurien, B. T. (2007, February 28). Awaken olfactory receptors of humans and experimental animals by coffee odourants to induce appetite. Retrieved from https://www.ncbi.nlm.nih.gov // Dorrier, J. (2015, January 25). If You Upload Your Mind to a Computer-Are You Still You? Retrieved from https://singularityhub.com // Dotinga, R. (2015, September 16). Brain Scans May Help Schizophrenia Treatment. Retrieved from http://www.webmd.com // Douglas, K., & Sutton, R. (2010, August 7). The Hidden Impact of Conspiracy Theories: Perceived and Actual Influence of Theories Surrounding the Death of Princess Diana. Retrieved from http://www.tandfonline.com // Dovey, D. (2014, October 06). Study Claims To Have Found Biological Cause For Underwear Fetishes. Retrieved from http://www.medicaldaily.com // Dovey, D. (2015, August 19). Researchers Claim They've Grown A Human Brain In A Lab. Retrieved from http://www.medicaldaily.com // Dovey, D. (2016, May 09). This Is Your Brain On 15 Languages. Retrieved from http://www.medicaldaily.com // Draganski, B., Gaser, C., Busch, V., & May, A. (2004, January 22). Neuroplasticity: Changes in grey matter induced by training. Retrieved from http://www.nature.com // Dreu, C. K., Greer, L. L., Handgraaf, M. J., Shalvi, S., Kleef, G. A., Baas, M., . . . Feith, S. W. (2010, June 11). The Neuropeptide Oxytocin Regulates Parochial Altruism in Intergroup Conflict Among Humans. Retrieved from http://science.sciencemag.org // Duke University. (2004, June 10). How Brain Gives Special Resonance To Emotional Memories. ScienceDaily. Retrieved from www.sciencedaily.com // Duncan, R. O. (2012, March 1). What Are the Structural Differences in the Brain between Animals That Are Self-Aware (Humans, Apes) and Other Vertebrates? Retrieved from https://www.scientificamerican.com // Dvorsky, G. (2012, September 03). Neuroscientists successfully control the dreams of rats. Could humans be next? Retrieved from http://io9.gizmodo.com // Dvorsky, G. (2013, April 04). The science behind meditation, and why it makes you feel better. Retrieved from http://io9.gizmodo.com // Dvorsky, G. (2013, January 09). The 12 cognitive biases that prevent you from being rational. Retrieved from http://io9.gizmodo.com // Dvorsky, G. (2013, September 26). The science behind power naps, and why they're so damn good for you. Retrieved from http://io9.gizmodo.com // Dvorsky, G. (2014, March 14). Our brains deliberately make us forget things, to prevent insanity. Retrieved from http://io9.gizmodo.com // Dvorsky, G. (2014, March 28). Why Asimov's Three Laws Of Robotics Can't Protect Us. Retrieved from http://io9.gizmodo.com // E. (2013, May 23). Sonic Science: The High-Frequency Hearing Test. Retrieved from https://www.scientificamerican.com // E. Aston. (2016, February 13). Brain-Computer Interface Lets You Control IoT Devices. Retrieved from https://wtvox.com // Editors. (2008, February 1). Why Do Some People Sleepwalk? Retrieved from http://www.scientificamerican.com // Edwards, S. Nightmares and the Brain. Retrieved from http://neuro.hms.harvard.edu // Effective Treatments for Anxiety. (2017, June 8). Retrieved from http://clinic.unc.edu // Ehrlich, B. (2014, December 03). Kanye West And Charli XCX Can See Music -- Wait, What? Retrieved from http://www.mtv.com // Ekman, P. (2015, November 11). Paul Ekman Group. Retrieved from http://www.paulekman.com // Elizabeth Hagen. (2009, November 4). Bats. ASU - Ask A Biologist. Retrieved from http://askabiologist.asu.edu // Emspak, J.

(2012, May 16). Brain-Controlled Robotic Arm Points Way to New Prosthetics. Retrieved from https://www.livescience.com // Entis, L. (2017, March 30). Depression Is the World's Most Widespread Illness. Retrieved from http://fortune.com // Epstein, R. (2017, June 27). Your brain does not process information and it is not a computer – Robert Epstein | Aeon Essays. Retrieved from https://aeon.co // Ericson, J. (2014, February 05). Boost Hearing By Spending Some Time In The Dark. Retrieved from http://www.medicaldaily.com // Erlacher, D. (2014, July 31). Practicing in Dreams Can Improve Your Performance. Retrieved from https://hbr.org // Esposito, M. (2013, May 11). Teaching the Brain to Calm Itself. Retrieved from http://brainblogger.com // Eveleth, R. (2015, July 21). Future - 'My robot bought illegal drugs' Retrieved from http://www.bbc.com // Everts, S. (2012, March 01). The Truth About Pheromones. Retrieved from http://www.smithsonianmag.com // Evidence of a 'higher' state of consciousness? (2017, April 19). Retrieved from https://www.sciencedaily.com // Fabello, A. M. (2013, January 18). The Neurobiology Behind All of the Ridiculous Things You Do When You're in Love. Retrieved from http://everydayfeminism.com // Fader, J. (2014, June 08). Diagnosing Game of Thrones: What's Ailing Westeros? Retrieved from https://www.psychologytoday.com // Falk, D. (2012, September 19). Could the Internet Ever "Wake Up"? Retrieved from http://www.slate.com // Fan, S. (2017, March 31). Animal Brains Networked Into Organic Computer 'Brainet' Retrieved from https://singularityhub.com // Fang, J. (2016, May 12). Running Can Boost Your Brain Power. Retrieved from http://www.iflscience.com // Fellizar, K. (2015, July 24). Millennials Are Masturbating To 'My Little Pony' Porn, Says New Survey, And Rainbow Dash Is The Fan Favorite. Retrieved from https://www.bustle.com // Fels, A. (2017, April 14). The Point of Hate. Retrieved from https://www.nytimes.com // Ferro, S. (2016, January 28). Want to Lucid Dream? Hit the Snooze Button. Retrieved from http://mentalfloss.com // Fessler, L. (2017, March 31). People who talk to pets, plants, and cars are actually totally normal, according to science. Retrieved from https://qz.com // Fetters, A. (2014, April 25). How to Achieve a Runner's High. Retrieved from http://www.runnersworld.com // Fields, H. (2012, April). Fragrant Flashbacks. Retrieved from http://www.psychologicalscience.org // Fields, R. D. (2011, November 16). Surety Bond: Breast-Feeding May Increase Children's IQ. Retrieved from https://www.scientificamerican.com // FindLaw. What is the Uniform Declaration of Death Act or UDDA. Retrieved from http://healthcare.findlaw.com // Fisher, H. (2008, July). Retrieved from https://www.ted.com // Flatow, I. (2013, October 25). Uncovering the Brain of a Psychopath. Retrieved from http://www.npr.org // Foer, J. (2006, April 27). No one has a photographic memory. Retrieved from http://www.slate.com // Fox, D. (2015, December). The Physical Limits to Genius. Retrieved from https://www.scientificamerican.com // Fox, M. (2013, June 19). This is your brain on happy: Machine can read your emotions. Retrieved from http://www.nbcnews.com // Fralick, M., Thiruchelvam, D., & Tien, H. C. (2016, February 08). Michael Fralick. Retrieved from http://www.cmaj.ca // Frank, A. (2016, February 9). A Mammal's Brain Has Been Cryonically Preserved and Recovered. Retrieved from https://motherboard.vice.com // Frank, C. (2017, February 1). Intelligence. Retrieved from https://www.psychologytoday.com // Frank, M. E., & Hettinger, T. P. (2005, January 01). What the Tongue Tells the Brain about Taste. Retrieved from https://academic.oup.com // Fung, B. (2012, September 17). When You Can't Sleep, How Good Is Lying in Bed With Your Eyes Closed? Retrieved from https://www.theatlantic.com // Furnham, A. (2014, February 24). Why Do People Join Cults? Retrieved from https://www.psychologytoday.com // G. (2017, March 30). "Depression: let's talk" says WHO, as depression tops list of causes of ill health. Retrieved from http://www.who.int // Gallagher, J. (2016, February 16). Doctors 3D-print 'living' body parts. Retrieved from http://www.bbc.com // Gammon, K. (2014, October 29). What Makes A Child Prodigy? Retrieved from http://www.popsci.com // Gasde, I. (1997). Cult experience : abuse, psychological distress, close relationships, and personality characteristics. Retrieved from http://scholarworks.montana.edu // GCSE Bitesize: Understanding memory. (2014). Retrieved from http://www.bbc.co.uk // Geddes, L. (2015, June 26). Future - The mystery of the female orgasm. Retrieved from http://www.bbc.com // Geek, B. (1970, January 01). The Brain Geek. Retrieved from http://thebraingeek.blogspot.com // Geggel, L. (2017, April 14). Painting a Song: Lorde's Synesthesia Turns Colors into Music. Retrieved from https://www.livescience.com // Gelb, I. J. (2016, July 07). Sumerian language. Retrieved from https://www.britannica.com // George the giant lobster liberated from restaurant. (2009, January 10). Retrieved from http://www.cnn.com // Gholipour, B. (2013, September 18). From the Deepest Coma, New Brain Activity Found. Retrieved from https://www.livescience.com // Ghose, T. (2013, February 26). Why Pain Can Feel Good. Retrieved from https://www.livescience.com // Ghose, T. (2014, January 16). 'Sixth Sense' Can Be Explained by Science. Retrieved from https://www.livescience.com // Ghose, T. (2014, March 25). Surprise! The Subconscious Mind Is Super Lie Detector. Retrieved from https://www.livescience.com // Gibbons, G. (2013, November 12). How Is a Stroke Diagnosed? Retrieved from https://www.nhlbi.nih.gov // Gleveckas-Martens, N. (2016, October 28). Somatosensory System Anatomy. Retrieved from http://emedicine.medscape.com // Goad, K. Goad, K. What Does Bipolar Mania Look and Feel Like? Retrieved from http://www.webmd.com Godman, H. (2014, April 9). Regular exercise changes the brain to improve memory, thinking skills. Retrieved from http://www.health.harvard.edu // Goertzel, T. (1994). Belief in Conspiracy Theories. Political Psychology, 15(4), 731-742. doi:10.2307/3791630 // Goetzke, K. (2014, June 24). Depression and Anxiety: An International Perspective. Retrieved from http://www.huffingtonpost.com // Goila, A. K., & Pawar, M. (2009). The diagnosis of brain death. Indian Journal of Critical Care Medicine : Peer-Reviewed, Official Publication of Indian Society of Critical Care Medicine, 13(1), 7–11. http://doi.org/10.4103/0972-5229.53108 // Goldhill, O. (2016, July 23). The "Jennifer Aniston neuron" is the foundation of compelling new memory research. Retrieved from https://qz.com // Goldin-Meadow, S. (1996). Kanzi: The Ape at the Brink of the Human Mind. Retrieved from http://pubpages.unh.edu // Goldman, J. G. (2012, October 17). Future - Is language unique to humans? Retrieved from http://www.bbc.com // Goldman, J. G. (2014, April 25). Future - What do animals dream about? Retrieved from http://www.bbc.com // Goleman, D. (1985, September 23). LEFT VS. RIGHT: BRAIN FUNCTION TIED TO HORMONE IN THE WOMB. Retrieved from http://www.nytimes.com // Goleman, D. (1988, May 16). Lies Can Point to Mental Disorders or Signal Normal Growth. Retrieved from http://www.nytimes.com // Goleman, D. (1995, March 27). The Brain Manages Happiness And Sadness in Different Centers. Retrieved from http://www.nytimes.com // Gómez-Pinilla, F. (2008, July). Brain foods: the effects of nutrients on brain function. Retrieved from https://www.ncbi.nlm.nih.gov // Gonzalez, R. (2012, March 14). The 4 Biggest Myths About the Human Brain. Retrieved from http://io9.gizmodo.com // Goodfriend, W. (2012, May 08). Classical Conditioning in "A Clockwork Orange". Retrieved from https://www.psychologytoday.com // Goodman, J. (2014, Fall). The Wages of Sinistrality: Handedness, Brain Structure, and Human Capital Accumulation. Retrieved from https://www.aeaweb.org // Gordon, B. (2013, January 1). Does Photographic Memory Exist? Retrieved from https://www.scientificamerican.com // Gorman, J. (2012, January 23). Survival's Ick Factor. Retrieved from http://www.nytimes.com // Gorman, T. E., & Green, C. S. (2016, April 18). Short-term mindfulness intervention reduces the negative attentional effects associated with heavy media multitasking. Retrieved from https://www.nature.com // Gosline, A. (2005, June 22). Why your brain has a 'Jennifer Aniston cell'. Retrieved from https://www.newscientist.com // Grady, D. (1993, June 1). July/August 2017. Retrieved from http://discovermagazine.com // Grant, A. (2014, January 02). The Dark Side of Emotional Intelligence. Retrieved from https://www.theatlantic.com // Grayling, A. C. (2011, June 22). Psychology: How we form beliefs. Retrieved from https://www.nature.com // Greenberg, G. (2006, September 01). Back From the Dead. Retrieved from https://www.wired.com // Greenberg, G. (2014, January 15). Lights Out: A New Reckoning for Brain Death. Retrieved from http://www.newyorker.com // Greenemeier, L. (2011, January 12). What Causes Someone to Act on Violent Impulses and Commit Murder? Retrieved from https://www.scientificamerican.com // Greenring, D. S. (2015, March 2). Can You Hear This Sound That Only Young People Can Hear? Retrieved from https://www.buzzfeed.com // Gregersen, P. K., Kowalsky, E., Lee, A., Baron-Cohen, S., Fisher, S. E., Asher, J. E., . . . Li, W. (2013, May 15). Absolute pitch exhibits phenotypic and genetic overlap with synesthesia. Retrieved from https://www.ncbi.nlm.nih.gov // Gregoire, C. (2015, January 20). How Our Sense Of Touch Affects Everything We Do. Retrieved from http://www.huffingtonpost.com // Gregoire, C. (2015, July 24). 5 Scary Health Effects Of Sleep Deprivation During The Teen Years. Retrieved from http://www.huffingtonpost.com // Gregoire, C. (2016, April 16). Consciousness Works Differently Than You Think, According To New Theory. Retrieved from http://www.huffingtonpost.com // Griffiths, J. (2016, April 13). This is your brain on LSD, literally. Retrieved from http://www.cnn.com // Grohol, J. M. (2009, October 26). Why 'Sleeping on It' Helps. Retrieved from https://www.livescience.com // Gronseth, G., & Greer, D. (2010). Life Alliance Organ Recovery Agency. Retrieved from http://surgery.med.miami.edu // Gross, R. E. (2015, June 05). Did the FDA Really Just Approve the "Female Viagra"? Not So Fast. Retrieved from http://www.slate.com // Guo, J. (2016, May 20). We're really good at forgetting all the terrible things we've done. Retrieved from https://www.washingtonpost.com // Hadhazy, A. (2011, December 30). Tip of the Tongue: Humans May Taste at Least 6 Flavors. Retrieved from https://www.livescience.com // Hagerty, B. B. (2009, May 20). Prayer May Reshape Your Brain ... And Your Reality. Retrieved from http://www.npr.org // Haidt, J. (2012, December). Retrieved from https://www.ted.com // Hall, H. (2010, January 26). Faith Healing. Retrieved from https://sciencebasedmedicine.org // Hall, J. (2016, May 04). Researchers identify autism-linked receptor that initiates synaptic pruning. Retrieved from https://www.extremetech.com // Hambling, D. (2001, January 31). A gleam in the eye. Retrieved from https://www.theguardian.com // Hambrick, D. Z. (2014, December 2). Brain Training Doesn't Make You Smarter. Retrieved from https://www.scientificamerican.com // Hamilton, J. (2006, March 29). Study Makes Case for Late Bloomers. Retrieved from http://www.npr.org // Hamilton, J. (2008, October 02). Think You're Multitasking? Think Again. Retrieved from http://www.npr.org // Hamilton, J. (2009, March 09). To The Brain, God Is Just Another Guy. Retrieved from http://www.npr.org // Hamilton, J. (2009, March 20). Smart People Really Do Think Faster. Retrieved from http://www.npr.org // Hamilton, J. (2012, June 04). What's Different About The Brains Of People With Autism? Retrieved from http://www.npr.org // Hamilton, K. (2016, August 15). Scientists Put A Worm's Mind Into A Robot's Body. Retrieved from http://www.iflscience.com // Hampton, D. (2016, March 23). How Your Thoughts Change Your Brain, Cells and Genes. Retrieved from http://www.huffingtonpost.com // Hanson, R. (1970, July 07). Emotion in the Brain. Retrieved from http://www.rickhanson.net // Harmon, K. (2009, March 13). God on the brain? Scientists map religious thoughts with scans. Retrieved from https://blogs.scientificamerican.com // Harmon, K. (2011, January 31). Aerobic exercise bulks up hippocampus, improving memory in older adults. Retrieved from https://blogs.scientificamerican.com // Harms, W. (2013, June 11). Perfect pitch may not be absolute after all. Retrieved from http://news.uchicago.edu // Harrington, R. (2015, September 30). Here's why all fast-food signs are red. Retrieved from http://www.businessinsider.com // Hartmann, E. (2009). Why do we dream? Retrieved from https://www.scientificamerican.com // Harvard Medical School. (2007, December). Why Do We Sleep, Anyway? Retrieved from http://healthysleep.med.harvard.edu // Harvard Medical School. Glossary. Retrieved from http://healthysleep.med.harvard.edu // Haselton, M., Nettle, D., & Murray, D. (2014, December 12). The Evolution of Cognitive Bias. Retrieved from http://www.sscnet.ucla.edu // Hayasaki, E. (2013, November 18). How Many of Your Memories Are Fake? Retrieved from https://www.theatlantic.com // Hayden, E. C. (2016, August 18). 'Radically rewritten' bacterial genome unveiled. Retrieved from http://www.nature.com // Hayes, L. (2013, January 28). How to Make Your Baby Smarter. Retrieved from http://www.huffingtonpost.com // Healy, M. (2012, June 05). That guilt you feel? There's a place (in your brain) for that. Retrieved from http://articles.latimes.com // Heid, M. (2014, June 18). The Science Behind Your Sex Fetish. Retrieved from http://www.shape.com // Heid, M. (2016, April 28). How Soon After an Orgasm Can You Start Again? Retrieved from http://www.menshealth.com // Hendrickson, J. Why Is it Impossible to Stop Thinking, to Render the Mind a Complete Blank? Retrieved from https://www.scientificamerican.com // Hendriksen, E. (2015, December 23). How to Stop Nightmares and Night Terrors. Retrieved from https://www.scientificamerican.com // Henry, A. (2013, November 20). What Anxiety Does to Your Brain and What You Can Do About It. Retrieved from http://lifehacker.com // Herculano-Houzel, S. (2009, November 9). The Human Brain in Numbers: A Linearly Scaled-up Primate Brain. Retrieved from https://www.ncbi.nlm.nih.gov // Herr, H. (2014, March). Retrieved from https://www.ted.com // Hickok, G. (2014, August 01). Opinion | Three Myths About the Brain. Retrieved from https://www.nytimes.com // Higher Perspective. (2016, May 09). New Research Says Overthinking Worriers Are Probably Creative Geniuses. Retrieved from http://www.higherperspectives.com // Hiles, D. (2001). Savant Syndrome. Retrieved from http://www.psy.dmu.ac.uk // Hiskey, D. (2013, August 20). What Happens When You Stick Your Head Into a Particle Accelerator. Retrieved from http://gizmodo.com // Hitti, M. (2005, January 31). Emotions Make the Memory Last. Retrieved from http://www.webmd.com // Hive minds: Honeybee intelligence creates a buzz. (2012, November 21). Retrieved from https://www.newscientist.com // Ho, V. (2016, July 28). Scientists in Singapore grow functioning 'mini' midbrain tissue. Retrieved from http://mashable.com // Hof, R. D. (2016, March 29). Is Artificial Intelligence Finally Coming into Its Own? Retrieved from https://www.technologyreview.com // Hogan, D. (2013, October 29). Eye tracking technology suggests people 'check out' women at first glance. Retrieved from https://www.sciencedaily.com // Hogan, D. (2014, July 21). Try, try again? Study says no: Trying harder makes it more difficult to learn some aspects of language, neuroscientists find. Retrieved from https://www.sciencedaily.com // Hogan, D. (2015, January 07). Two brain regions join forces for absolute pitch. Retrieved from https://www.sciencedaily.com // Hogan, D. (2016, February 23). New study finds our desire for 'like-minded others' is hard-wired. Retrieved from https://www.sciencedaily.com // Hogan, D. (2017, April 19). Evidence of a 'higher' state of consciousness? Retrieved from https://www.sciencedaily.com // Hogan, D. (2017, June 15). Why do those with autism avoid eye contact? Retrieved from https://www.sciencedaily.com // Hogan, D. (2017, May 08). Cannabis reverses aging processes in the brain, study suggests. Retrieved from https://www.sciencedaily.com // Horgan, J. Can a Single Brain Cell "Think"? If So, What Does That Imply about the "Neural Code"? Retrieved from https://blogs.scientificamerican.com // Horowitz, A. (2013, December 01). Why Brain Size Doesn't Correlate With Intelligence. Retrieved from http://www.smithsonianmag.com // Horowitz, K. (2016, March 07). The Brains of Anxious People May Perceive the World Differently. Retrieved from http://mentalfloss.com // Howard, J. (2015, May 29). Yes! Acquiring 'Perfect' Pitch Is Possible For Some Adults, Scientists Say. Retrieved from http://www.huffingtonpost.com // Hu, J. C. (2014, August 20). The Strange World of Koko, Kanzi, and the Decline of Ape Language Research. Retrieved from http://www.slate.com // Hughes, J. (2014, October 14). Ancient memory palace provides classical connection to 2014 Nobel prize in medicine. Retrieved from http://theconversation.com // Hullinger, J. (2015, February 17). 7 Things We Can Turn Off and On in the Brain. Retrieved from http://mentalfloss.com // Hurley, D. (2014, January 13). How Sex Affects Intelligence, and Vice Versa. Retrieved from http://www.theatlantic.com // Hurley, D. (2015, November 17). The Return of Electroshock Therapy. Retrieved from https://www.theatlantic.com // Hutchinson, S. (2013, September 03). Is It True That Elephants Never Forget? Retrieved from http://mentalfloss.com // Hyman, I. E. (2014, April 28). The Dangers of Going on Autopilot. Retrieved from https://www.psychologytoday.com // HyperPhysics. Sensitivity of the Human Ear. Retrieved from http://hyperphysics.phy-astr.gsu.edu // I love Rock 'n' Roll'-Music genre preference modulates brain responses to music. (2013, February). Retrieved from http://www.sciencedirect.com // IBM Research Deep Blue Overview. (2001, February 23). Retrieved from https://www.research.ibm.com // Images, D. Z. (2017, February 8). Compulsive Behavior Isn't Necessarily A Sign of A Broken Brain. Retrieved from http://nymag.com // Indivero, V. M. (2014, November 12). Learning languages is a workout for brains, both young and old. Retrieved from http://news.psu.edu // Inglis-Arkell, E. (2011, July 23). Adults may be better at learning languages than children. Retrieved from http://io9.gizmodo.com // Ingmire, J. (2015, May 28). Acquiring 'perfect' pitch may be possible for some adults. Retrieved from https://news.uchicago.edu // Ingraham, C. (2016, April 25). Heavy pot use can permanently damage short-term memory, study shows. Retrieved from http://www.seattletimes.com // Insel, T. (2011, August 12). Post by Former NIMH Director Thomas Insel: Mental Illness Defined as Disruption in Neural Circuits. Retrieved from https://www.nimh.nih.gov // Insel, T. (2011, December 06). Post by Former NIMH Director Thomas Insel: Antidepressants: A complicated picture. Retrieved from https://www.nimh.nih.gov // Jabr, F. (2012, November 1). Self-Awareness with a Simple Brain. Retrieved from https://www.scientificamerican.com // Jabr, F. (2013, March 20). Let's Get Physical: The Psychology of Effective Workout Music. Retrieved from https://www.scientificamerican.com // Jabr, F. (2016, January 1). How Sugar and Fat Trick the Brain into Wanting More Food. Retrieved from https://www.scientificamerican.com // Jackson Senior Research Fellow in the School of Health Sciences, RMIT University, M., & Schembri Post-doctoral research fellow, School of Health Sciences, RMIT University, R. (2015, August 12). Unravelling the mysteries of sleep: how the brain 'sees' dreams. Retrieved from http://theconversation.com // Jackson, J. (2013, February 22). What Happens To Your Brain When You Get Black-Out Drunk? Retrieved from http://gizmodo.com // Jackson, M. (2015). Creativity and Psychotic States in Exceptional People. Retrieved from https://books.google.com // Jacobson, R. (2015, April 01). Memories May Not Live in Neurons' Synapses. Retrieved from https://www.scientificamerican.com // Jacoby, S. (2014, June 21). 10 Famous Geniuses With Truly Weird Secret Habits. Retrieved from http://listverse.com // Jaeger, K. (2016, April 24). The Crazy Thing Your Brain Does When You Sleep in a New Place. Retrieved from https://www.attn.com // Jaffe, E. (2015, May 19). Morning People Vs. Night Owls: 9 Insights Backed By Science. Retrieved from https://www.fastcodesign.com // Jarrett, C. (2015, August 19). The Neuroscience of Being a Selfish Jerk. Retrieved from http://nymag.com // Jarrett, C. (2015, November 5). When Jealousy and Empathy Collide in the Brain. Retrieved from http://nymag.com // Jarrett, C. (2017, June 03). A Calm Look at the Most Hyped Concept in Neuroscience - Mirror Neurons. Retrieved from https://www.wired.com // Jha, A. (2005, June 30). Where belief is born. Retrieved from https://www.theguardian.com // Jha, A. (2013, August 28). Miniature brains grown in test tubes – a new path for neuroscience? Retrieved from https://www.theguardian.com // Johns Hopkins Medicine. (2016, May 19). Fruit fly brains shed light on why we get tired when we stay up too late. Retrieved from https://www.sciencedaily.com // Johnson, D. (2013, July 17). Music in the Brain: The Mysterious Power of Music. Retrieved from http://dujs.dartmouth.edu // Jones, B. (2014, October 20). Kleine-Levin Syndrome: Why the rare 'Sleeping Beauty' illness is no fairytale. Retrieved from http://www.bbc.com // Judis, J. B. (2014, October 25). Are Political Beliefs Predetermined at Birth? Retrieved from https://newrepublic.com // Kalat, J. W. (2012, January 1). Biological Psychology. Retrieved from https://books.google.com // Kalb, C. (2013, January 01). Your Alarm Clock May Be Hazardous to Your Health. Retrieved from http://www.smithsonianmag.com // Kamer, F. (2015, July 13). 3 (Scientifically Proven) Ways to Learn a New Language. Retrieved from http://mentalfloss.com // Kannan, M. (2015). Some Rules of Language are Wired in the Brain.

Retrieved from https://www.scientificamerican.com // Kasey Yturralde. (2009, September 30). Did You Know Butterflies are Legally Blind? . ASU - Ask A Biologist. Retrieved from http://askabiologist.asu.edu // Katigbak, R. (2014, April 24). Heroin Is the Most Dangerous Way to Increase Your Creativity. Retrieved from https://www.vice.com // Kaufman, S. B. (2014, February 25). Where do Savant Skills Come From? Retrieved from https://blogs.scientificamerican.com // Kaufman, S. B. (2014, February 3). What Do IQ Tests Test?: Interview with Psychologist W. Joel Schneider. Retrieved from https://blogs.scientificamerican.com // Kayser, M. (2015, July 28). Sleepy Fruitflies Get Mellow: Sleep Deprivation Reduces Aggression, Mating Behavior in Flies, Penn Study Finds – PR News. Retrieved from https://www.pennmedicine.org // Kean, S. (2014, May 06). The True Story of Phineas Gage Is Much More Fascinating Than the Mythical Textbook Accounts. Retrieved from http://www.slate.com // Kershenbaum, A., Bowles, A. E., Freeberg, T. M., Jin, D. Z., Lameira, A. R., & Bohn, K. (2014, October 07). Animal vocal sequences: not the Markov chains we thought they were. Retrieved from http://rspb.royalsocietypublishing.org // Keyser, H. (2015, October 22). Do You Use Only 10% of Your Brain? Retrieved from http://mentalfloss.com // Keysers, C. (2013, July 24). Inside the Mind of a Psychopath – Empathic, But Not Always. Retrieved from https://www.psychologytoday.com // Khan Academy. Autonomic nervous system (ANS) and physiologic markers of emotion. Retrieved from https://www.khanacademy.org // Khan Academy. Emotions: cerebral hemispheres and prefrontal cortex. Retrieved from https://www.khanacademy.org // Khan Academy. Emotions: limbic system. Retrieved from https://www.khanacademy.org // Khan, A. (2014, October 29). Liberal or conservative? Brain's 'disgust' reaction holds the answer. Retrieved from http://www.latimes.com // Khan, A. (2016, February 3). Scientists 3D-print a 'brain' to learn the secret behind its folds. Retrieved from http://www.latimes.com // Khan, K. S., & Chaudhry, S. (2015, February 09). An evidence-based approach to an ancient pursuit: systematic review on converting online contact into a first date. Retrieved from http://ebm.bmj.com // Khatchadourian, R. (2015, January 19). We Know How You Feel. Retrieved from http://www.newyorker.com // Khazan, O. (2015, February 24). How Smartphones Hurt Sleep. Retrieved from https://www.theatlantic.com // Kihlstrom, J. F. (2017, January 3). Memory. Retrieved from http://socrates.berkeley.edu // Kim, M. (2014, August 22). Chirps, whistles, clicks: Do any animals have a true 'language'? Retrieved from https://www.washingtonpost.com // Kloc, J. (2011, November 1). Putting Insomnia on Ice. Retrieved from https://www.scientificamerican.com // KLS Foundation. What is KLS? Retrieved from http://klsfoundation.org // Kluger, J. (2015, November 9). The Science Behind Pets That Find Their Way Home. Retrieved from http://time.com // Knight, W. (2016, July 08). Self-Driving Cars Get a Code of Ethics. Retrieved from https://www.technologyreview.com // Knight, W. (2016, March 21). Who Will Choose AI's Ethical Code? Retrieved from https://www.technologyreview.com // Koch, C. (2011, November 1). Probing the Unconscious Mind. Retrieved from https://www.scientificamerican.com // Koch, C. (2016, January 1). Does Size Matter--for Brains? Retrieved from https://www.scientificamerican.com // Koerth-baker, M. (2013, May 25). Why Rational People Buy Into Conspiracy Theories. Retrieved from http://www.nytimes.com // Kohn, M. (2014, July 29). Future - The truth about smart drugs. Retrieved from http://www.bbc.com // Kolbert, E. (2017, February 27). Why Facts Don't Change Our Minds. Retrieved from http://www.newyorker.com // Konnikova, M. (2014, January 11). Opinion | Goodnight. Sleep Clean. Retrieved from https://www.nytimes.com // Konnikova, M. (2015, March 4). The Power of Touch. Retrieved from http://www.newyorker.com // Koren, M. (2013, July 15). Wait, Have I Been Here Before? The Curious Case of Déjà Vu. Retrieved from http://www.smithsonianmag.com // Kotler, S. (2014, July 21). The Uncanniest Valley: What Happens When Robots Know Us Better Than We Know Ourselves? Retrieved from https://www.forbes.com // Kotler, S. (2015, March 23). The Feel-Good Switch: The Radical Future of Emotion. Retrieved from https://singularityhub.com // Kremer, W. (2015, April 18). The strange afterlife of Einstein's brain. Retrieved from http://www.bbc.com // Kuruvilla, C. (2016, May 09). Can This Interactive Map Of Emotions Bring World Peace? The Dalai Lama Thinks So. Retrieved from http://www.huffingtonpost.com // Kuszewski, A. (2011, March 7). You Can Increase Your Intelligence: 5 Ways to Maximize Your Cognitive Potential. Retrieved from https://blogs.scientificamerican.com // Kwan, N. (2016, April 5). Cold blast treatment relieves phantom limb pain, study finds. Retrieved from http://www.foxnews.com // Kwon, D. (2015, November 24). What Makes Our Brains Special? Retrieved from https://www.scientificamerican.com // Kwon, D. (2017). Are Prions behind All Neurodegenerative Diseases? Retrieved from https://www.scientificamerican.com // LaBarre, S. (2012, July 30). How a Collar Could Help Deaf People. Retrieved from https://www.fastcompany.com // LaFrance, A. (2015, March 20). About That Breastfeeding Study. Retrieved from https://www.theatlantic.com // Lahl, O. (2008, February 11). An ultra short episode of sleep is sufficient to promote declarative memory performance. Retrieved from http://onlinelibrary.wiley.com // Lallanilla , M. (2014, January 08). A Pill For Perfect Pitch? New Research Suggests There May Be One. Retrieved from http://www.huffingtonpost.com // LaMotte, S. (2016, January 08). No, you haven't read this déjà vu story before. Retrieved from http://www.cnn.com // Langley, L. (2013, January 15). 10 Interesting Facts About Your Brain on Sex. Retrieved from http://www.alternet.org // Langley, L. (2015, September 08). Do Animals Dream? Retrieved from http://news.nationalgeographic.com // Langone, M. D. (2013, September 29). How Cults Rewire the Brain. Retrieved from http://www.huffingtonpost.com // Lavelle, J. (2015, October 8). New Brain Effects behind "Runner's High". Retrieved from https://www.scientificamerican.com // Lavelle, M. (2014, September 22). Mothers May Pass Lyme Disease to Children in the Womb. Retrieved from https://www.scientificamerican.com // Lebowitz, S. (2015, July 31). Scientists say chronic stress can actually change your brain. Retrieved from http://www.businessinsider.com // Lehrer, J. (2009, April 20). Magic and the Brain: Teller Reveals the Neuroscience of Illusion. Retrieved from https://www.wired.com // Lehrer, J. (2017, June 19). Why Smart People Are Stupid. Retrieved from http://www.newyorker.com // Lemonick, M. D. (2007, August 09). Explaining Déjà Vu. Retrieved from http://content.time.com // Lenzen, M. (2013). Feeling Our Emotions. Retrieved from https://www.scientificamerican.com // Leu, C. (2015, October 12). Scientists Can Now Predict Intelligence From Brain Activity. Retrieved from https://www.wired.com // Lewis, J. G. (2012, August 14). The Neuroscience of Déjà Vu. Retrieved from https://www.psychologytoday.com // Lewis, J. G. (2015, January 12). Smells Ring Bells: How Smell Triggers Memories and Emotions. Retrieved from https://www.psychologytoday.com // Lewis, M. (2016, December 14). Why it's high time that attitudes to addiction changed. Retrieved from https://aeon.co // Lewis, T. (2013, August 20). Delusional People See the World Through Their Mind's Eye. Retrieved from https://www.livescience.com // Lewis, T. (2013, August 27). Humans Can Learn to Echolocate. Retrieved from https://www.livescience.com // Lewis, T. (2013, March 05). New Theory Explains Why Amputees Feel Phantom Pain. Retrieved from http://www.livescience.com // Lewis, T. (2013, May 01). Brain Region Found to Control Aging. Retrieved from https://www.livescience.com // Lewis, T. (2014, February 10). Galileo's Optical Illusion Explained by Neuroscience. Retrieved from https://www.livescience.com // Lewis, T. (2014, February 14). 5 Ways Love Affects the Brain. Retrieved from https://www.livescience.com // Lewis, T. (2015, April 06). Near-Death Experiences: What Happens in the Brain Before Dying. Retrieved from https://www.livescience.com // Lewis, T. (2015, August 20). Here's what the brain of an extremely selfish person looks like. Retrieved from http://www.businessinsider.com // Lewis, T. (2016, March 25). Human Brain: Facts, Functions & Anatomy. Retrieved from http://www.livescience.com // Liljas, P. (2014, June 8). Computer Passes Turing Test, Achieves AI Milestone. Retrieved from http://time.com // Linden, D. J. (2015, March 20). The Neurobiology of BDSM Sexual Practice. Retrieved from https://www.psychologytoday.com // Livio, M. (2014, January 15). Brains of Genius. Retrieved from http://www.huffingtonpost.com // Loria, K. (2014, April 23). Brain Implants Could Give People Perfect Memories And Night Vision. Retrieved from http://www.businessinsider.com // Loui, P., Zamm, A., & Schlaug, G. (2013, March 13). Absolute Pitch and Synesthesia: Two Sides of the Same Coin? Shared and Distinct Neural Substrates of Music Listening. Retrieved from https://www.ncbi.nlm.nih.gov // Lubinski, D. (2015). The Probability of Genius. Retrieved from http://www.slate.com // Lucid Dream State Staff. Why Do Reality Tests Work in Dreams? Retrieved from http://theluciddreamsite.com // Lund University. (2012, October 08). Language learning makes the brain grow, Swedish study suggests. Retrieved from https://www.sciencedaily.com // Luntz, S. (2016, August 15). Why Cheaters Don't Prosper. Retrieved from http://www.iflscience.com // M. (2017, May 23). Probiotic use linked to improved symptoms of depression. Retrieved from https://www.sciencedaily.com // M. M. (2011, February 22). Why Speaking More than One Language May Delay Alzheimer's. Retrieved from http://healthland.time.com // M.D., P. J. (2011, March 07). Music, Rhythm and The Brain » Brain World. Retrieved from http://brainworldmagazine.com // Mackey, A. (2014, September 04). What happens in the brain when you learn a language? Retrieved from https://www.theguardian.com // Magazine, C. P. (2013, May 23). An Itch Is Not a Low-Level Form of Pain. Retrieved from https://www.scientificamerican.com // Mailonline, A. H. (2017, April 05). Haunting faces of the Bedlam ladies: Portraits show patients at asylum where 'problem' women were dumped by families even if they were sane.. and paying gawkers watched their cruel 'treatment' . Retrieved from http://www.dailymail.co.uk // MailOnline, L. E. (2010, January 26). Who REALLY needs more sleep - men or women? One of Britain's leading sleep experts says he has the answer. Retrieved from http://www.dailymail.co.uk // Main, D. (2013, May 06). Iron Balls in Bird Brains May Sense Magnetic Fields. Retrieved from https://www.livescience.com // Maines, R. (2017, June 27). Vibrators and hysteria: how a cure became a female sexual icon. Retrieved from http://theconversation.com // Malick, A. (1970, August 29). Humans Emit Sex Scent Signals. Retrieved from http://abcnews.go.com // Manza, L. (2016, April 14). How Cults Exploit One of Our Most Basic Psychological Urges. Retrieved from http://www.huffingtonpost.com // Marder, J. (2015, February 27). That dress isn't blue or gold because color doesn't exist. Retrieved from http://www.pbs.org // Marlborough, P. (2016, October 9). What It's Like to Have Borderline Personality Disorder. Retrieved from https://www.vice.com // Maron, D. F. (2015, September 1). Mediterranean Eating Habits Prove Good for the Brain. Retrieved from https://www.scientificamerican.com // Marr, B. (2016, April 15). The Future of Sex: Sensors, Cognitive Computing and Robotics. Retrieved from https://www.forbes.com // Martin, L. J. (2016, August 22). Aging changes in the senses. Retrieved from https://medlineplus.gov // Martinez-Conde, S. (2015, January 17). Out of Mind, Out of Sight: Suppressed Unwanted Memories Are Harder to See. Retrieved from https://blogs.scientificamerican.com // Mastin, L. (2013). TYPES AND STAGES OF SLEEP. Retrieved from https://www.howsleepworks.com // Max Planck Society. (2016, April 28). Study shows reptiles share REM and slow-wave sleep patterns with mammals, birds. Retrieved from https://phys.org // Mayo Clinic Staff. (2015, August 14). Coma. Retrieved from http://www.mayoclinic.org // Mayo Clinic Staff. (2015, September 01). Narcolepsy. Retrieved from http://www.mayoclinic.org // Mayo Clinic Staff. (2017, February 15). Bipolar disorder. Retrieved from http://www.mayoclinic.org // McAuliffe, K. (2011, January 20). July/August 2017. Retrieved from http://discovermagazine.com // Mccarthy, E. (2014, July 16). Why Do We Laugh When We're Tickled? Retrieved from http://mentalfloss.com // McFarland, M. (2016, July 14). 300-pound mall robot runs over toddler. Retrieved from http://money.cnn.com // McGreevey, S. (2011, January 21). Mindfulness meditation training changes brain structure in 8 weeks. Retrieved from http://www.massgeneral.org // McKie, R. (2012, January 28). Cloning scientists create human brain cells. Retrieved from https://www.theguardian.com // McLaughlin, B. Myths and Misconceptions about Second Language Learning: What Every Teacher Needs to Unlearn. Retrieved from https://people.ucsc.edu // McLeod, S. (2007). Saul McLeod. Retrieved from https://www.simplypsychology.org // McNeely, E., Gale, S., Tager, I., Kincl, L., Bradley, J., Coull, B., & Hecker, S. (2014, March 10). The self-reported health of U.S. flight attendants compared to the general population. Retrieved from https://ehjournal.biomedcentral.com // McPhate, M. (2016, June 10). United States of Paranoia: They See Gangs of Stalkers. Retrieved from https://www.nytimes.com // MD, C. C. (2014, June 30). "Social Jet Lag" and the Teenager - Craig Canapari, MD -. Retrieved from http://drcraigcanapari.com // Meeusen, R. (2014). Exercise, Nutrition and the Brain. Sports Medicine (Auckland, N.z.), 44(Suppl 1), 47–56. http://doi.org/10.1007/s40279-014-0150-5 // Melatonin - Overview. (2016, October 21). Retrieved from http://www.webmd.com // Melina, R. (2011, January 12). What's the Difference Between the Right Brain and Left Brain? Retrieved from https://www.livescience.com // Mennella, J. A., Jagnow, C. P., & Beauchamp, G. K. (2001, June). Prenatal and postnatal flavor learning by human infants. Retrieved from https://www.ncbi.nlm.nih.gov // Meredith, M. (2001, May). Human vomeronasal organ function: a critical review of best and worst cases. Retrieved from https://www.ncbi.nlm.nih.gov // Merritt, A. (2013, September 18). Are children really better at foreign language learning? Retrieved from http://www.telegraph.co.uk // Metz, C. (2017, June 03). IBM Dreams Impossible Dream With Clone of Human Brain. Retrieved from https://www.wired.com // Miller, A. (2014, February). The criminal mind. Retrieved from http://www.apa.org // Miller, G. (2017, June 03). 9 Things to Know About Reviving the Recently Dead. Retrieved from https://www.wired.com // Miller, K. (2016, November 04). The Number of People Who've Slept with a Coworker Will Astound You. Retrieved from http://www.womenshealthmag.com // Miller, R. (2015, December 02). Big Data Still Requires Humans To Make Meaningful Connections. Retrieved from https://techcrunch.com // Miller, R. (2016, July 17). Too much big data running through my brain. Retrieved from https://techcrunch.com // Miller, S. G. (2015, October 27). The Spooky Effects of Sleep Deprivation. Retrieved from https://www.livescience.com // Miller, S. G. (2016, February 01). What's That Word? Marijuana May Affect Verbal Memory. Retrieved from https://www.livescience.com // Misaki, M., Suzuki, H., Savitz, J., Drevets, W. C., & Bodurka, J. (2016, February 16). Individual Variations in Nucleus Accumbens Responses Associated with Major Depressive Disorder Symptoms. Retrieved from https://www.nature.com // Mistysyn, Mark. (2017, February 06). Using 3D Printing Technology to Print Organs and Tissue. Retrieved from http://www.wakehealth.edu // Monks, K. (2015, October 20). Training the brain to push the body beyond its limits. Retrieved from http://www.cnn.com // Monroe, D. (2004, November 9). Speaking Tonal Languages Promotes Perfect Pitch. Retrieved from http://www.scientificamerican.com // Montero-Luque, C. (2014, November 05). Humans Can Make the Internet of Things Smarter. Retrieved from https://hbr.org // Mooallem, J. (2016, April 26). 'Are We Smart Enough to Know How Smart Animals Are?' and 'The Genius of Birds' Retrieved from https://www.nytimes.com // Mooney, C. (2011, September 07). Your Brain on Politics: The Cognitive Neuroscience of Liberals and Conservatives. Retrieved from http://blogs.discovermagazine.com // More about Pigs: The Humane Society of the United States. (2012, September 26). Retrieved from http://www.humanesociety.or // Morrison, J. (2003). Frida Kahlo. Retrieved from https://books.google.com // Moskowitz, C. (2011, March 04). Criminal Minds Are Different From Yours, Brain Scans Reveal. Retrieved from https://www.livescience.com // Mossop, B. (2011, March 11). Elite Athletes' Brains Bigger in Some Areas. Retrieved from https://www.wired.com // Moyer, M. W. (2012, September 24). Fish Oil Supplement Research Remains Murky. Retrieved from https://www.scientificamerican.com // MPG. (2013, June 25). Hunger affects decision making and perception of risk. Retrieved from https://www.mpg.de // Mulcahy, C. (2016, April 27). A Math Genius Like No Other Comes to the Big Screen. Retrieved from https://blogs.scientificamerican.com // Murphy, M. (2015, September 11). This mind-controlled prosthetic robot arm lets you actually feel what it touches. Retrieved from https://qz.com // Muse, T. (2016, December 18). 19 Things to Try When You Can't Sleep – Thrive Global. Retrieved from https://journal.thriveglobal.com // Mustafa, B., Evrim, Ö, & Sarı, A. (2005). Secondary Mania Following Traumatic Brain Injury. The Journal of Neuropsychiatry and Clinical Neurosciences, 17(1), 122-124. doi:10.1176/jnp.17.1.122 // Myer, G. D., Yuan, W., D., B. F., Smith, D., Altaye, M., Reches, A., . . . Krueger, D. (2016, April 29). The Effects of External Jugular Compression Applied during Head Impact Exposure on Longitudinal Changes in Brain Neuroanatomical and Neurophysiological Biomarkers: A Preliminary Investigation. Retrieved from http://journal.frontiersin.org // Myer, G. D., Yuan, W., Foss, K. D., Thomas, S., Smith, D., Leach, J., . . . Altaye, M. (2016, June 15). Analysis of head impact exposure and brain microstructure response in a season-long application of a jugular vein compression collar: a prospective, neuroimaging investigation in American football. Retrieved from http://bjsm.bmj.com // N. (2017, February 5). NAMI: National Alliance on Mental Illness. Retrieved from https://www.nami.org // Nagel, T. (1974). What is it like to be a bat? Retrieved from http://organizations.utep.edu // National Center for Complementary and Integrative Health. Meditation: In Depth. (2017, March 17). Retrieved from https://nccih.nih.gov // National Institute of Health. (2017, January 19). Brain Research through Advancing Innovative Neurotechnologies (BRAIN). Retrieved from https://www.braininitiative.nih.gov // National Institute on Aging. (2012, September). The Search for Alzheimer's Prevention Strategies. Retrieved from https://www.nia.nih.gov // National Science Foundation - Where Discoveries Begin. (2010, June 28). Retrieved from https://www.nsf.gov // National Sleep Foundation. Retrieved from https://sleepfoundation.org // National Sleep Foundation. Animals' Sleep: Is There a Human Connection? (2003). Retrieved from https://sleepfoundation.org // National Sleep Foundation. Narcolepsy. Retrieved from https://sleepfoundation.org // National Sleep Foundation. REM Sleep Behavior Disorder. Retrieved from https://sleepfoundation.org // National Sleep Foundation. What is Insomnia? Retrieved from https://sleepfoundation.org // National Stroke Association. What is stroke? (2016, March 16). Retrieved from http://www.stroke.org // Nauert, R. (2016, June 28). "Herd" Mentality Explained. Retrieved from https://psychcentral.com // Naziri, J. (2016, July 12). The future of listening to music is feeling it, according to Timbaland. Retrieved from https://techcrunch.com // Neith, K. (2014, January 30).Worry on the Brain | Caltech. Retrieved from http://www.caltech.edu // Neuroscience for Kids. "Oh Say Can You Say". Retrieved from https://faculty.washington.edu // Neuroscience for Kids. Amazing Animal Senses. Retrieved from https://faculty.washington.edu // Neuroscience for Kids. How Much Do Animals Sleep? Retrieved from https://faculty.washington.edu // Neuroskeptic. (2016, May 28). A Recurring Sickness: Pathological Déjà Vu. Retrieved from http://blogs.discovermagazine.com // Newberg, A. (2011, May 31). Religious Experiences Shrink Part of the Brain. Retrieved from https://www.scientificamerican.com // Newell, B. (2008, August 11). Complex decision? Don't sleep on it. Retrieved from https://www.eurekalert.org // News Staff. (2014, August 26). Is ESP Real? Harvard Scientists Say They Have Settled The Debate. Retrieved from http://www.science20.com // Nicholson, M. (2015, December 23). Can groups of people "remember" something that didn't happen? Retrieved from http://www.hopesandfears.com // Nielsen, J. A., Zielinski, B. A., Ferguson, M. A., Lainhart, J. E., & Anderson, J. S. (2013, August 14). An Evaluation of the Left-Brain vs. Right-Brain Hypothesis with Resting State Functional Connectivity Magnetic Resonance Imaging. Retrieved from http://journals.plos.org // NIH. Alzheimer's Disease Fact Sheet. (2011, August 17). Retrieved from https://www.nia.nih.gov // NIH. Aneurysm | MedlinePlus. Retrieved from https://medlineplus.gov // NIH. Schizophrenia. (2016, February). Retrieved from https://www.nimh.nih.gov // NIH. Science Capsule - Cochlear Implants. (2016, September 26). Retrieved from https://www.nidcd.nih.gov // NIH. Technology and the Future of Mental Health Treatment. Retrieved from https://www.nimh.nih.gov // NIH. The Teen Brain: 6 Things to Know. Retrieved from https://www.nimh.nih.gov // Nordqvist,

C. (2017, January 21). Amnesia: Causes, Symptoms, and Treatments. Retrieved from http://www.medicalnewstoday.com // Novella, S. (2015, May 20). The PIED Piper of Nootropics. Retrieved from https://sciencebasedmedicine.org // Noyes, K. (2016, June 06). A former NASA chief just launched this AI startup to turbocharge neural computing. Retrieved from http://www.pcworld.com // NSF. (2015, November 2). National Science Foundation - Where Discoveries Begin. Retrieved from https://www.nsf.gov // Nutt, D. (2012, June 10). Is the future of drugs safe and non-addictive? | David Nutt. Retrieved from https://www.theguardian.com // NYU Langone Medical Center. (2009, January 13). Delusions Associated With Consistent Pattern Of Brain Injury. Retrieved from https://www.sciencedaily.com // O. (2016, March 8). The Science of Turning Her On. Retrieved from http://www.wnyc.org // Obsessive Compulsive Disorder Among Adults. Retrieved from https://www.nimh.nih.gov // O'Connor, P. (2014, May 01). Is Sex Addiction Real? Retrieved from https://www.psychologytoday.com // Office of Public and Intergovernmental Affairs. (2016, July 7). Office of Public and Intergovernmental Affairs. Retrieved from https://www.va.gov // Olewitz, C. (2016, April 28). Incredible DIY synesthesia mask gives you the ability to smell colors — without taking drugs. Retrieved from http://www.digitaltrends.com // Olson, J. (2013, January 1). How to Prevent Jet Lag. Retrieved from http://www.scientificamerican.com // Osborne, C. (2016, June 13). MIT's artificial intelligence passes key Turing test. Retrieved from http://www.zdnet.com // Ossola, A. Inside the Brain of a Super Memorizer. Retrieved from https://braindecoder.com // P., Baker, D., MRCVS, D. P., Gray, A., Rohrbacher, J., Smith, M., . . . Banks, T. (2016, June 30). Why Are Cats and Dogs Color-Blind? Retrieved from http://www.petful.com // Palermo, E. (2014, January 29). What Happens When You Die? Retrieved from https://www.livescience.com // Palermo, E. (2015, August 21). New Robotic Exoskeleton Is Controlled by Human Thoughts. Retrieved from https://www.livescience.com // Palmer, K. M. (2017, June 06). 'Brain Balls' Grown From Skin Cells Spark With Electricity. Retrieved from https://www.wired.com // Pappas, S. (2011, February 17). Top 10 Controversial Psychiatric Disorders. Retrieved from https://www.livescience.com // Pappas, S. (2014, February 19). The New Yoga? Sadomasochism Leads to Altered States, Study Finds. Retrieved from https://www.livescience.com // Parish, D. (2016, February 19). Lyme: The Infectious Disease Equivalent of Cancer, Says Top Duke Oncologist. Retrieved from http://www.huffingtonpost.com // Park, M. Y. (2014, March 14). How Our Sense of Taste Changes as We Age. Retrieved from http://www.bonappetit.com // Parker, C. B. (2014, May 20). 'Tiger moms' vs. Western-style mothers? Stanford researchers find different but equally effective styles. Retrieved from http://news.stanford.edu // Parrish, A. (2015, June 17). Female Viagra: Here's What It's Like. Retrieved from http://time.com // Parry, W. (2010, December 05). Now You See It: Neuroscientists Reveal Magicians' Secrets. Retrieved from https://www.livescience.com // Patterson, F. (1980, October 9). Gorilla Talk. Retrieved from http://www.nybooks.com // Patterson, F., Washburn, S., & Gardner, M. (1981, April 2). More On Ape Talk. Retrieved from http://www.nybooks.com // Pedersen, T. (2015, October 06). Meditation Shown to Alter Gray Matter in Brain. Retrieved from https://psychcentral.com // Peever, J. What Happens in the Brain During Sleep? Retrieved from https://www.scientificamerican.com // Penenberg, A. L. (2012, September 07). NeuroFocus Uses Neuromarketing To Hack Your Brain. Retrieved from https://www.fastcompany.com // Perry, C. J. (2013, October 29). Are animals as smart, or as dumb, as we think they are? Retrieved from https://phys.org // Pham, P. (2015, August 28). The Impacts Of Big Data That You May Not Have Heard Of. Retrieved from https://www.forbes.com // Phung, A. (2007, February 04). Behavioral Finance: Key Concepts - Herd Behavior. Retrieved from http://www.investopedia.com // Picciuto, E. (2015, February 25). They Don't Want an Autism Cure. Retrieved from http://www.thedailybeast.com // Pilley, John. (2013, August 27). Is this the world's cleverest dog? Retrieved from http://www.telegraph.co.uk // Pincott, J. (2012, March 13). Slips of the Tongue. Retrieved from https://www.psychologytoday.com // Pinkowski, J. (2015, November 22). Scientists Pinpoint Where Happiness Lives in the Brain. Retrieved from http://mentalfloss.com // Pinkowski, J. (2015, October 25). Tracing the Evolution of the Human Brain Through Casts of the Inner Skull. Retrieved from http://mentalfloss.com // Pinsker, J. (2014, October 01). The Psychology Behind Costco's Free Samples. Retrieved from https://www.theatlantic.com // Poirier, V. (2016, August 10). 7 Reasons Beauty And The Beast Is Not a Tale Of Stockholm Syndrome. Retrieved from http://thefederalist.com // Pollack, J., & Cabane, O. F. (2016, May 11). Your Brain Has A "Delete" Button - Here's How to Use It. Retrieved from https://www.fastcompany.com // Poltrack, E. What Processes in the Brain Allow You to Remember Dreams? Retrieved from https://www.scientificamerican.com // Popova, M. (2015, September 17). Are We Nearing the Maximum Capacity of the Human Brain? Retrieved from https://www.brainpickings.org // Popular Science. Mantis Shrimp Vision Is Not As Mindblowing As You've Been Told. (2014, June 19). Retrieved from http://www.popsci.com // Porkka-Heiskanen, T., & Kalinchuk, A. V. (2011, April). Adenosine, energy metabolism and sleep homeostasis. Retrieved from https://www.ncbi.nlm.nih.gov // Porter, J. (2014, August 27). How To Rewire Your Brain For Greater Happiness. Retrieved from https://www.fastcompany.com // Pranav Mistry. Sixthsense. Retrieved from http://www.pranavmistry.com // Psychology Today. Creativity. Retrieved from https://www.psychologytoday.com // Publications, H. H. (2009, June). What causes depression? Retrieved from http://www.health.harvard.edu // Publications, H. H. (2012, August). Boost your memory by eating right. Retrieved from http://www.health.harvard.edu // Publications, H. H. Forgetfulness - 7 types of normal memory problems. Retrieved from http://www.health.harvard.edu // Purdy, K. (2010, July 13). What Caffeine Actually Does to Your Brain. Retrieved from http://lifehacker.com // Purdy, M. C. (2015, April 23). Extra sleep fixes memory problems in flies with Alzheimer's-like condition | The Source | Washington University in St. Louis. Retrieved from https://source.wustl.edu // Purves, D. (1970, January 01). The Auditory Cortex. Retrieved from https://www.ncbi.nlm.nih.gov // Quiroga, R. Q. Searching for the Jennifer Aniston Neuron [Excerpt]. Retrieved from https://www.scientificamerican.com // Raine, A. (2013, April 26). The Criminal Mind. Retrieved from https://www.wsj.com // Raizen, D. (2007, August 21). Lethargus is a Caenorhabditis elegans sleep-like state. Retrieved from http://www.nature.com // Ramachandran, V. (2011, February 14). V.S. Ramachandran's Tales Of The 'Tell-Tale Brain' Retrieved from http://www.npr.org // Rapaport, L. (2015). Mediterranean Diet with Olive Oil, Nuts Linked to Healthier Brain. Retrieved from https://www.scientificamerican.com // Rathi , A. (2013, August 11). New meta-analysis checks the correlation between intelligence and faith. Retrieved from https://arstechnica.com // Ratliff, E. (2006, July 01). Déjà Vu, Again and Again. Retrieved from http://www.nytimes.com // Ratner, P. (2016, August 06). This Man Will Get the World's First Human Head Transplant Procedure. Retrieved from http://bigthink.com // Ravindran, S. (2015, January 20). Are we all born with a talent for synaesthesia? – Shruti Ravindran | Aeon Essays (P. Weintraub, Ed.). Retrieved from https://aeon.co // Ray, C. C. (2000, October 30). Brain Folds. Retrieved from http://www.nytimes.com // Reardon, S. (2015, September 28). How Your Brain Is Wired Reveals the Real You. Retrieved from https://www.scientificamerican.com // Reas, E. (2013, October 15). Important New Theory Explains Where Old Memories Go. Retrieved from http://www.scientificamerican.com // Reddy, S. (2013, September 02). The Perfect Nap: Sleeping Is a Mix of Art and Science. Retrieved from https://www.wsj.com // Reef, C. (2013, Summer). SUMMER 2013 CONTENTS. Retrieved from http://sm.stanford.edu // Reese | June 8, 2016, 8:52 AM PST, H. (2016, June 8). New research shows that Swarm AI makes more ethical decisions than individuals. Retrieved from http://www.techrepublic.com // Regis, E., & Church, G. (2012, October 17). July/August 2017. Retrieved from http://discovermagazine.com // Reilly, L. Why Do Our Best Ideas Come in the Shower? (2013, September 06). Retrieved from http://mentalfloss.com // Repertinger, S., Fitzgibbons, W. P., Omojola, M. F., & Brumback, R. A. (2006, July). Long survival following bacterial meningitis-associated brain destruction. Retrieved from https://www.ncbi.nlm.nih.gov // Research Institute of Molecular Pathology. (2012, October 23). The fabric for weaving memory: To establish long-term memory, neurons have to synthesize new proteins. ScienceDaily. Retrieved from www.sciencedaily.com // Resnick, B. (2016, March 18). If you're just not a morning person, science says you may never be. Retrieved from https://www.vox.com // Rhodes, J. (2013, June). Why Do I Think Better after I Exercise? Retrieved from https://www.scientificamerican.com // Ringo, A. (2013, August 09). Understanding Deafness: Not Everyone Wants to Be 'Fixed' Retrieved from https://www.theatlantic.com // Roach, J. (2010, June 24). Touching Heavy, Hard Objects Makes Us More Serious. Retrieved from http://news.nationalgeographic.com // Robison, E. Can We Control Our Thoughts? Why Do Thoughts Pop into My Head as I'm Trying to Fall Asleep? Retrieved from https://www.scientificamerican.com // Robson, D. (2011, March 30). Memory sticks: Do mnemonics work? Retrieved from https://www.newscientist.com // Robson, D. (2015, September 18). Future - Is another human living inside you? Retrieved from http://www.bbc.com // Rodriguez, T. (2013, May 1). Negative Emotions Are Key to Well-Being. Retrieved from https://www.scientificamerican.com // Rodriguez, T. (2013, November 19). Sex 'Addiction' Isn't a Guy Thing. Retrieved from https://www.theatlantic.com // Romano, N. (2011, July 13). Dreams About Falling: Dream Meanings Explained. Retrieved from http://www.huffingtonpost.com // Romm, C. (2014, August 20). Those Who Know They're Dreaming Tend to Be Savvier When Awake. Retrieved from https://www.theatlantic.com // Romm, C. (2015, February 13). This Is Your Brain on Magic. Retrieved from https://www.theatlantic.com // Ronson, J. (2011, May 27). The Psychopath Test. Retrieved from https://www.thisamericanlife.org // Rony, Paz. (2016, March 7). People With Anxiety Show Fundamental Differences in Perception. Retrieved from https://www.weizmann-usa.org // Rosenfeld, J. (2015, September 01). Two New Studies Explore the Neuroscience of Negative Emotions. Retrieved from http://mentalfloss.com // Rothstein, R. (2014, January 7). The Urban Poor Shall Inherit Poverty. Retrieved from http://prospect.org // Rougeau, M. (2012, June 27). Google IO 2012: Google introduces Siri-killer Google Now. Retrieved from http://www.techradar.com // Ruthsatz, J., & Urbach, J. (2012). Child prodigy: A novel cognitive profile places elevated general intelligence, except working memory and attention to detail, at the root of prodigiousness. Retrieved from http://scottbarrykaufman.com // Ryan, J. (2014, June 03). Are 'Tiger Moms' Better Than Cool Moms? Retrieved from https://www.theatlantic.com // Sacks, O. (2012, December 12). Seeing God in the Third Millennium. Retrieved from https://www.theatlantic.com // Sakai, J. (2012, March 15). A wandering mind reveals mental processes and priorities. Retrieved from http://news.wisc.edu // Sample, I. (2011, November 16). Brain scans of happy people help explain their 'rose-tinted' outlook. Retrieved from https://www.theguardian.com // Sanders, R. (2015, July 09). Jet lagged and forgetful? It's no coincidence. Retrieved from http://news.berkeley.edu // Sanders, R. (2016, September 20). Brain's hippocampus helps fill in the blanks of language. Retrieved from http://news.berkeley.edu // Sankar-Gorton, E. (2015, October 29). Retrieved from http://www.huffingtonpost.com // Sansone, R. A., & Sansone, L. A. (2010, May). Fatal Attraction Syndrome: Stalking Behavior and Borderline Personality. Retrieved from https://www.ncbi.nlm.nih.gov // Schachner, E. How Has the Human Brain Evolved? Retrieved from https://www.scientificamerican.com // Schjødt, U. (2012, May 02). The Neuroscience of Prayer. Retrieved from http://www.theeuropean-magazine.com // Schocker, L. (2014, January 08). Here's A Horrifying Picture Of What Sleep Loss Will Do To You. Retrieved from http://www.huffingtonpost.com // School of Life Sciences | Ask A Biologist. (2011, May 08). Retrieved from https://askabiologist.asu.edu // Schulte, B. (2015, May 26). Harvard neuroscientist: Meditation not only reduces stress, here's how it changes your brain. Retrieved from https://www.washingtonpost.com // Schulte, B. (2015, May 26). Harvard neuroscientist: Meditation not only reduces stress, here's how it changes your brain. Retrieved from https://www.washingtonpost.com // Scott, K. (2006, March 31). Sleepwalking chef's recipe for disaster. Retrieved from https://www.theguardian.com // Scutti, S. (2013, May 10). Genetics And Neurobiology: The Future Of Bipolar Disorder Treatment And Diagnosis. Retrieved from http://www.medicaldaily.com // Scutti, S. (2015, February 04). The Brain's Gray Matter Goes When Illness Is Around. Retrieved from http://www.medicaldaily.com // Seaberg, M. (2012, March 03). Synesthetes: "People of the Future". Retrieved from https://www.psychologytoday.com // Semedo, P. D. (2017, February 13). Alzheimer's May Be Delayed by Speaking a Second Language, Study Suggests. Retrieved from https://alzheimersnewstoday.com // Sergo, P. (2010, May 1). Going Out with a Bang. Retrieved from https://www.scientificamerican.com // Shaer, M. (2014, November 01). Is This the Future of Robotic Legs? Retrieved from http://www.smithsonianmag.com // Shah, A. (2013, December 6). A new disorder is plaguing teenagers: Sleep texting. Retrieved from http://www.startribune.com // Sharman, L., & Dingle, G. A. (2015). Extreme Metal Music and Anger Processing. Retrieved from https://www.ncbi.nlm.nih.gov // Shaw, J. (2016, March 14). How False Memory Changes What Happened Yesterday. Retrieved from https://blogs.scientificamerican.com // Sheikh, K. (2016, February 05). Early Bird or Night Owl? It May Be in Your Genes. Retrieved from https://www.livescience.com // Sheridan, L., & James, D. (2015, June 16). Complaints of group-stalking ('gang-stalking'): an exploratory study of their nature and impact on complainants. Retrieved from http://www.tandfonline.com // Shermer, M. (2014, December 1). Why Do People Believe in Conspiracy Theories? Retrieved from https://www.scientificamerican.com // Sifferlin, A. (2014, January 15). Mashed Up Memory: How Alcohol Speeds Memory Loss in Men. Retrieved from http://healthland.time.com // Silvestro, S. (2016, June 24). A New View of Phineas Gage. Retrieved from https://hms.harvard.edu // Singal, J. (2015, April 13). Researchers Found the 'Bystander Effect' in 5-Year-Olds. Retrieved from http://nymag.com // Singal, J. (2015, November 4). A New Study Suggests That Sleeping on a Decision Might Not Do Much. Retrieved from http://nymag.com // Sisley, D. (Ed.). (2015, July 24). The dA-Zed guide to Jean-Michel Basquiat. Retrieved from http://www.dazeddigital.com // Skomorowsky, A. (2015, March 10). How Molly Works in the Brain. Retrieved from https://www.scientificamerican.com // Sleep Drive and Your Body Clock. Retrieved from https://sleepfoundation.org // Small, G. (2010, September 28). Mass Hysteria Can Strike Anywhere, Anytime. Retrieved from https://www.psychologytoday.com // Smith, D. (2012, January 3). 50-million-year-old cricket and katydid fossils from Colorado hint at origin of insect hearing. Retrieved from http://www.colorado.edu // Snyder (3), T., & Gackenbach, J. (1988, January 01). Individual Differences Associated with Lucid Dreaming. Retrieved from https://link.springer.com // Society for Neuroscience. (2012, April 1). The Neuron. Retrieved from http://www.brainfacts.org // Solso, R. Cerebrum. (2004, July 01). Retrieved from http://dana.org // Soniak, M. (2012, January 05). How Do Magic Eye Pictures Work? Retrieved from http://mentalfloss.com // Sonnenburg, J. S. (2015, May 1). Gut Feelings–the "Second Brain" in Our Gastrointestinal Systems [Excerpt]. Retrieved from https://www.scientificamerican.com // Sparrow, B., Liu, J., & Wegner, D. M. (2011, August 05). Google Effects on Memory: Cognitive Consequences of Having Information at Our Fingertips. Retrieved from http://science.sciencemag.org // Spector, D. (2013, October 16). How Cats See The World Compared To Humans [PICTURES]. Retrieved from http://www.businessinsider.com // Spiegel, A. (2013, May 20). If Your Shrink Is A Bot, How Do You Respond? Retrieved from http://www.npr.org // Springer. (2015, June 23). Men think they are maths experts, therefore they are. Retrieved from https://www.sciencedaily.com // Staff, American Kennel Club. (2015, November 12). Are Dogs Color Blind? Retrieved from http://www.akc.org // Staff, NPR. (2014, May 22). Overexposed? Camera Phones Could Be Washing Out Our Memories. Retrieved from http://www.npr.org // Staff, Pagesix. (2016, May 24). The very, very, very strange life and times of Prince. Retrieved from http://pagesix.com // Staff, The American Cancer Society. (2015, April 10). Placebo Effect. Retrieved from https://www.cancer.org // Stafford, T. (2012, March 13). Future - Why can smells unlock forgotten memories? Retrieved from http://www.bbc.com // Starr, D. (2015, March 5). Remembering a Crime That You Didn't Commit. Retrieved from http://www.newyorker.com // Stetka, B. (2016, April 26). Do Vitamins and Supplements Make Antidepressants More Effective? Retrieved from https://www.scientificamerican.com // Stix, G. (2011, December 5). Will You Live Forever or until Your Next Software Release by Uploading Your Brain into a Computer? Retrieved from https://blogs.scientificamerican.com // Stringer, C. Why Have Our Brains Started to Shrink? Retrieved from https://www.scientificamerican.com // Ströbele, Jonathan. (2013). Major System database. Retrieved from http://major-system.info // Stromberg, J. (2012, April 03). The Benefits of Daydreaming. Retrieved from http://www.smithsonianmag.com // Sue, C. (2013, August 21). How Do Birds Navigate? Retrieved from https://www.nationalgeographic.org // Swain, F. (2014, September 24). Future - Cyborgs: The truth about human augmentation. Retrieved from http://www.bbc.com // Swaminathan, N. (2007, September 13). Fact or Fiction?: Babies Exposed to Classical Music End Up Smarter. Retrieved from https://www.scientificamerican.com // Swaminathan, N. The Fear Factor: When the Brain Decides It's Time to Scram. Retrieved from https://www.scientificamerican.com // Swenson, R. (2006). Chapter 11 - The Cerebral Cortex. Retrieved from http://www.dartmouth.edu // Szalavitz, M. (2012, July 10). What Genius and Autism Have in Common. Retrieved from http://healthland.time.com // Szalavitz, M. (2013, June 05). Sexual and Emotional Abuse Scar the Brain in Specific Ways. Retrieved from http://healthland.time.com // Szpunar, K. (2007, January 03). Memory And Future Thought Go 'Hand-In-Hand' Retrieved from https://www.medicalnewstoday.com // Tallinen, T., Chung, J. Y., Rousseau, F., Girard, N., Lefèvre, J., & Mahadevan, L. (2016, February 01). On the growth and form of cortical convolutions. Retrieved from http://www.nature.com // Tannenbaum, M. (2015, October 31). Avoiding Zombie Attacks with Social Psychology. Retrieved from https://blogs.scientificamerican.com // Taren, A. A., Creswell, J. D., & Gianaros, P. J. Dispositional Mindfulness Co-Varies with Smaller Amygdala and Caudate Volumes in Community Adults. Retrieved from http://journals.plos.org // Tasca, C., Rapetti, M., Carta, M. G., & Fadda, B. (2012). Women And Hysteria In The History Of Mental Health. Retrieved from https://www.ncbi.nlm.nih.gov // Tatera, K. (2016, April 20). 8 Tips to Craft the Perfect Tinder Profile, Based on Scientific Research. Retrieved from http://thescienceexplorer.com // Tatera, K. (2016, February 18). Brains over Beauty? Male Dating Preferences Are Evolving, Research Finds. Retrieved from http://thescienceexplorer.com // Taub, B. (2016, August 15). Sleepwalkers Who Injure Themselves Don't Feel Pain Until They Wake Up. Retrieved from http://www.iflscience.com // Taub, B. (2016, February 25). Scientists Successfully Erase Associative Memories In Mice. Retrieved from http://www.iflscience.com // Taub, B. (2016, March 17). Suppressing Memories May Cause Amnesia. Retrieved from http://www.iflscience.com // Taylor, E. (2016, June 02). 'The Witness' Looks Back At Those Accused Of Ignoring A Murder. Retrieved from http://www.npr.org // Taylor, K. (2013, November 20). The science of brainwashing. Retrieved from https://www.thenakedscientists.com // Teale , J. C., & O'Connor, A. R. (2015, March 3). What is Déjà Vu? Retrieved from https://blogs.scientificamerican.com // Terrace, H. (1980, December 4). More on Monkey Talk. Retrieved from http://www.nybooks.com // Than, K. (2005, February 22). Rare but Real: People Who Feel, Taste and Hear Color. Retrieved from http://www.livescience.com // Than, K. (2007, April 11). How Sight and Sound Can Trick Your Brain. Retrieved from https://www.livescience.com // The Dana Foundation. (2015, December 8). Big Data and the Brain: Peeking at the Future of Neuroscience. Retrieved from http://www.brainfacts.org // The Economist. Hacking your brain. (2015, March 05). Retrieved from http://www.economist.com // The Editors Encyclopedia Britannica. (2017, January 13). Retrieved from https://www.britannica.com // The Guardian. Did dinosaurs have really small brains? (2009, February 08). Retrieved from https://www.theguardian.com // The Hemispherectomy Foundation. Facts About Hemispherectomy. (2009). Retrieved from http://hemifoundation. homestead.com // The New York Times. Questions Answered: Invented Languages. (2010, March 10). Retrieved from https://schott.blogs.nytimes.com // The School of Life. Developing Emotional Intelligence. Retrieved from http://www.thebookoflife.org // The Science Behind Eyewitness Identification Reform. Retrieved from https://www.innocenceproject.org // Think Again: Men and Women Share Cognitive Skills. (2014, August). Retrieved from www.apa.org // Thomas, B. The Limits of Fight-or-Flight Training. (2016, January 05). Retrieved from http://blogs.discovermagazine.com // Thomson, H. (2014, July 2). Consciousness on-off switch discovered deep in brain. Retrieved from https://www.newscientist.com // Toohey, P. (2014). Jealousy. Retrieved from https://books.google.com // Trafton, A. (2015, December 16). Music in the brain. Retrieved from http://news.mit.edu // Trafton, A. (2015, March 31). How the brain processes emotions. Retrieved from http://news.mit.edu // Trouble at the lab. (2013, October 18). Retrieved from http://www. economist.com // Troxell, M. Schopenhauer, Arthur Internet Encyclopedia of Philosophy. Retrieved from http:// www.iep.utm.edu // Tsoulis-Reay, A. (2015, February 26). What It's Like to See 100 Million Colors. Retrieved from http://nymag.com // Tufekci, Z. (2015, August 10). Opinion | Why 'Smart' Objects May Be a Dumb Idea. Retrieved from https://www.nytimes.com // Turner, R. How to Make Lucid Dreams Last Longer. Retrieved from http://www.world-of-lucid-dreaming.com // Twilley, N. (2015, November 2). The Illusion of Taste. Retrieved from http://www.newyorker.com // Twomey, S. (2010, January 01). Phineas Gage: Neuroscience's Most Famous Patient. Retrieved from http://www.smithsonianmag.com // U.S. National Library of Medicine. Cerebellum - function. (2016, January 5). Retrieved from https://medlineplus.gov // UCSC. What is a dream? How is the word "dream" defined? Retrieved from http://www2.ucsc.edu // UCSF Memory and Aging Center. What is Alzheimer's disease? Retrieved from http://memory.ucsf.edu // UCSF. Brain 101: Topics in Neuroscience. Retrieved from http://memory.ucsf.edu // University of California - Los Angeles. (2007, December 13). Different Areas Of The Brain Respond To Belief, Disbelief And Uncertainty. ScienceDaily. Retrieved June 30, 2017 from www.sciencedaily.com // University of Cambridge. The deluded brain. Retrieved from http://www. neuroscience.cam.ac.uk // University of Lincoln. (2012, November 15). Key to super-sensory hearing? Newly identified hearing organ in bushcrickets' ears may inspire acoustic sensors. ScienceDaily. Retrieved from www.sciencedaily.com // USCSF. Brain 101: Topics in Neuroscience. Retrieved from http://memory.ucsf.edu // Venosa, A. (2015, July 18). Scientists Create Mini Brains To Better Understand Autism. Retrieved from http:// www.medicaldaily.com // Victora, C., Horta, B. L., De Mola, C., Quevedo, L., Pinhiero, R., Gigante, D., . . . Barros, F. (2015, April). Association between breastfeeding and intelligence, educational attainment, and income at 30 years of age: a prospective birth cohort study from Brazil. Retrieved from http://www.thelancet. com // Viegas, J. (2010, January 22). Dolphin Intelligence Explained. Retrieved from https://www.seeker.com // Viegas, J. (2014, April 02). 10 Surprising Facts About Animal Intelligence. Retrieved from https://www.seeker. com // Vince, G. (2006, February 16). 'Sleeping on it' best for complex decisions. Retrieved from https://www. newscientist.com/decisions/ Vincent, J. (2016, June 22). Google's AI researchers say these are the five key problems for robot safety. Retrieved from https://www.theverge.com // Vincent, J. (2016, March 24). Twitter taught Microsoft's friendly AI chatbot to be a racist asshole in less than a day. Retrieved from https://www. theverge.com // Vogel , E. K., & Drew, T. (2008, November 04). Why Do We Forget Things? Retrieved from https://www.scientificamerican.com // Vokov, N. (2014, April). Preface. Retrieved from https://www. drugabuse.gov // Wade, N. (2017, June 01). You Look Familiar. Now Scientists Know Why. Retrieved from https://www.nytimes.com // Wai, J. (2016, February 08). How Psychologically Well Adjusted Are Gifted People? Retrieved from https://www.psychologytoday.com // Walcutt, D. L. (2016, July 17). Stages of Sleep. Retrieved from https://psychcentral.com // Waller, J. (2008, September 17). John Waller on the mystery of mass hysteria. Retrieved from https://www.theguardian.com // Walton, A. G. (2015, February 09). 7 Ways Meditation Can Actually Change The Brain. Retrieved from https://www.forbes.com // Wanjek, C. (2006, August 29). The Tongue Map: Tasteless Myth Debunked. Retrieved from http://www.livescience.com // Wanjek, C. (2013, September 03). Left Brain vs. Right: It's a Myth, Research Finds. Retrieved from https://www. livescience.com // Wanjek, C. (2015, January 28). Psychopaths' Brains Don't Grasp Punishment, Scans Reveal. Retrieved from https://www.livescience.com // Wanjek, C. (2015, November 11). Brain Scan May Predict Chance of Coma Recovery. Retrieved from https://www.livescience.com // Wanshel, E. (2016, May 19). This Earpiece Translates Foreign Languages For You In Real Time. Retrieved from http://www.huffingtonpost.com // Ward, B. (2013, March 23). In Anorexia Nervosa, Brain Responds Differently to Hunger Signals. Retrieved from https://health.ucsd.edu // Ward, D. M. (2013, December 1). The Internet Has Become the External Hard Drive for Our Memories. Retrieved from https://www.scientificamerican.com // Watson, C. J., Baghdoyan, H. A., & Lydic, R. (2010, December). Neuropharmacology of Sleep and Wakefulness. Retrieved from https://www. ncbi.nlm.nih.gov // Watts, A. (2014, January 1). Why Do We Develop Certain Irrational Phobias? Retrieved from https://www.scientificamerican.com // Wcisel , M. (2012, May 1). Sharkwatch SA Blog. Retrieved from http:// www.sharkwatchsa.com // WebMD. (2000, March 06). Get Smart: Brain Cells Do Regrow, Study Confirms. Retrieved from http://www.webmd.com // Wein, H. (Ed.). (2013, October 28). How Sleep Clears the Brain. Retrieved from https://www.nih.gov // Weir, K. (2012, June). The roots of mental illness. Retrieved from http:// www.apa.org // Weir, K. (2017, May). Why we believe alternate facts. Retrieved from http://www.apa.org // Weller, C. (2014, July 10). This Audio Illusion Reveals Your Brain's Knack For Language. Retrieved from http:// www.medicaldaily.com // Weller, C. (2014, June 18). Behind The Scenes Of Your Fight-or-Flight Response. Retrieved from http://www.medicaldaily.com // Weller, C. (2015, May 21). Kool-Aid, Advertising, And The Science Of Everyday Brainwashing. Retrieved from http://www.medicaldaily.com // Welsh, J. (2012, July 01). Human yawns unleash dog yawns. Retrieved from http://nbcnews.com // Welsh, J. (2012, June 22). How Cocaine Vaccines Could Cure Drug Addiction. Retrieved from https://www.livescience.com // Wengenroth, M., Blatow, M., Heinecke, A., Reinhardt, J., Stippich, C., Hofmann, E., & Schneider, P. (2014, May). Increased volume and function of right auditory cortex as a marker for absolute pitch. Retrieved from https://www.ncbi. nlm.nih.gov // Wenk, G. L. (2015, November 14). This Is Why You Wanted Coffee and Donuts This Morning. Retrieved from https://www.psychologytoday.com // West, R. F., Meserve, R. J., & Stanovich, K. E. (2012, September). Cognitive sophistication does not attenuate the bias blind spot. Retrieved from https://www.ncbi. nlm.nih.gov // Westcott, K. (2013, August 22). What is Stockholm syndrome? Retrieved from http://www.bbc. com // Wheeler, M. (2009, March 17). Study gives more proof that intelligence is largely inherited. Retrieved from http://newsroom.ucla.edu // Wheeling, K. (2016, January 11). The brains of men and women aren't really that different, study finds. Retrieved from http://www.sciencemag.org // Whitbourne, S. K. (2011, October 18). Your Smartphone May Be Making You... Not Smart. Retrieved from https://www.psychologytoday.com // Whitsett, D. (2014, November 1). Why Cults Are Harmful Neurobiological Speculations. Retrieved from http:// www.icsahome.com // Wighton, K. (2016, July 8). How research on LSD revealed: first scans show how the drug affects the brain. Retrieved from http://www3.imperial.ac.uk // Wilcox, S. (2016, September 14). In Your Dreams. Retrieved from https://sleepfoundation.org // Wiley-Blackwell. (2009, March 27). Brain Activity Associated With Phantom Limbs, Study Shows. Retrieved from https://www.sciencedaily.com // Wilford, J. N. (2007, April 16). Almost Human, and Sometimes Smarter. Retrieved from http://www.nytimes.com // Willingham, A. (2016, May 05). Was Iwo Jima flag-raising a false memory? Retrieved from http://www.cnn.com // Wilson, D How do we manage to remember smells despite the fact that each olfactory sensory neuron only survives for about 60 days and is then replaced by a new cell? Retrieved from https://www.scientificamerican. com // Wired Staff. (2011, November 29). Digital Narcotics May Be the Future of Drugs. Retrieved from https:// www.wired.com // Wittmann, M., Dinich, J., Merrow, M., & Roenneberg, T. (2006). Social jetlag: misalignment of biological and social time. Retrieved from https://www.ncbi.nlm.nih.gov // Wlassoff, V. (2015, January 24). How Does Post-Traumatic Stress Disorder Change the Brain? Retrieved from http://brainblogger.com // Wlassoff, V. How the Brain Recognizes Faces. Retrieved from http://brainblogger.com // Wolchover, N. (2012, June 02). Why Are Genius and Madness Connected? Retrieved from https://www.livescience.com // Wolchover, N. (2012, September 18). What Will Future Humans Look Like? Retrieved from https://www.livescience.com // Wolford, B. (2014, July 08). For People With Bipolar Disorder, Risky Behavior Is Like Crack. Retrieved from http://www.medicaldaily.com // Wong, K. Explaining the Sense of Touch. Retrieved from https://www. scientificamerican.com // Yahoo Health. (2014, July 29). 5 Brain Myths That Won't Go Away. Retrieved from https://www.yahoo.com // Yong, E. (2011, January 24). Self-control in childhood predicts health and wealth in adulthood. Retrieved from http://phenomena.nationalgeographic.com // Yong, E. (2012, March 23). Future - Will we ever... talk to the animals? Retrieved from http://www.bbc.com // Yong, E. (2014, April 21). This is How You Study The Evolution of Animal Intelligence. Retrieved from http://phenomena.nationalgeographic.com // Yong, E. (2015, August 27). How Reliable Are Psychology Studies? Retrieved from https://www.theatlantic.com // Yuhas, D. (2013, March 1). Is Cocoa the Brain Drug of the Future? Retrieved from http://www. scientificamerican.com // Yuhas, D. (2014, May 1). Are Human Pheromones Real? Retrieved from https://www. scientificamerican.com // Zang, S., Shmader, T., & Hall, W. M. (2016, July 30). Women's true math skills

unlocked by pretending to be someone else. Retrieved from https://digest.bps.org.uk // Zetter, K. (2017, June 03). Brinks' Super-Secure Smart Safes: Not So Secure. Retrieved from https://www.wired.com // Zimmer, C. (2009, September 10). The Brain: Where Does Sex Live in the Brain? From Top to Bottom. Retrieved from http://discovermagazine.com // Zimmer, C. (2010, April 16). July/August 2017. Retrieved from http:// discovermagazine.com

PHOTO CREDITS

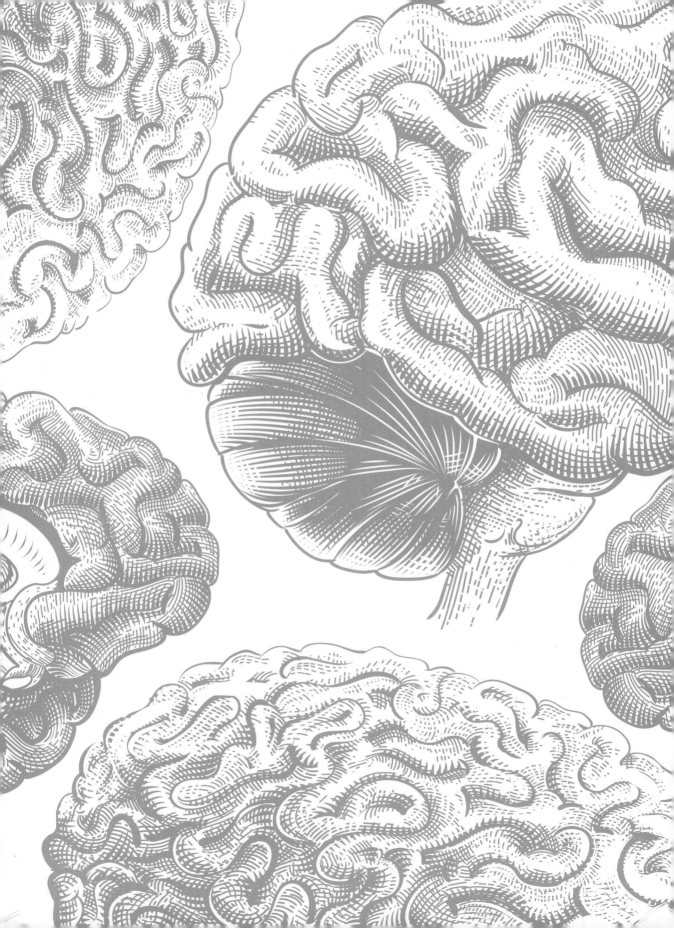